THE POLITICAL ECONOMY OF NATURAL GAS

Natural gas is a key commodity in the world economy and industrial and domestic consumers are increasingly dependent on it. This is increasingly affecting and changing international relations between importer and supplier countries. The Siberian natural gas pipeline which supplies Soviet gas to Western Europe is a good example of the impact of natural gas on international relations. This book presents a comprehensive survey of the world natural gas industry: it looks at the problems of supply, the pattern of demand, the economics of the industry and how the industry is being affected by changes in other energy sectors. The book concludes with an assessment of future prospects.

Ferdinand E. Banks, Associate Research Professor at the University of Uppsala, Sweden and Visiting Professor at the University of Stockholm.

CROOM HELM COMMODITY SERIES
Edited by Fiona Gordon-Ashworth, Bank of England

URANIUM: A STRATEGIC SOURCE OF ENERGY
Marian Radetzki, Institute for International Economic Studies, Stockholm

TIN: ITS PRODUCTION AND MARKETING
William Robertson, University of Liverpool

INTERNATIONAL COMMODITY CONTROL:
A Contemporary History and Appraisal
Fiona Gordon-Ashworth, Bank of England

COMMODITY MODELS FOR FORECASTING AND POLICY ANALYSIS
Walter C. Labys and Peter K. Pollak

THE MODERN PLANTATION IN THE THIRD WORLD
Edgar Graham with Ingrid Floering

THE INTERNATIONAL GRAIN TRADE: PROBLEMS AND PROSPECTS
Nick Butler

The series is edited by Fiona Gordon-Ashworth, formerly of the University of Southampton, who now works at the Bank of England. (The views expressed in this book are not to be taken as those of the Bank of England.)

The
Political Economy of
Natural Gas

Ferdinand E. Banks

CROOM HELM
London • New York • Sydney

© 1987 Ferdinand Banks
Croom Helm Ltd, Provident House, Burrell Row,
Beckenham, Kent, BR3 1AT

Croom Helm Australia, 44-50 Waterloo Road,
North Ryde, 2113, New South Wales

Published in the USA by
Croom Helm
in association with Methuen, Inc.
29 West 35th Street
New York, NY 10001

British Library Cataloguing in Publication Data

Banks, Ferdinand E.
  The political economy of natural gas. —
  (Croom Helm commodity series)
  1. Gas, Natural — Political aspects 2. Gas
  industry — Political aspects
  3. International economic relations
  I. Title
  338.2'7285   HD9581
  ISBN 0-7099-3940-X

Library of Congress Cataloging-in-Publication Data

ISBN 0-7099-3940-X

Photosetting by Mayhew Typesetting, Bristol, England
Printed and bound in Great Britain
by Billing & Sons Limited, Worcester.

Contents

List of Figures
List of Tables
Preface

1	**Introduction: Geology, Units and Background**	1
	Foreword	2
	Geology and history	3
	Units and conversion factors	6
	Reserves, production, use and trade in natural gas	9
2	**Energy Trends**	16
	Outlook for oil	16
	World steam coal market	22
	Natural gas	31
	Appendix: reserve/production ratio	39
3	**Economic Theory and Natural Gas**	41
	Liquefied natural gas	47
	Offshore gas	55
4	**Natural Gas in the United States**	58
	Supply, price and demand of natural gas	59
	Spot market for natural gas	64
	Regulation and deregulation	66
	Canada and Mexico	71
	Appendix: prices and deregulation	74
5	**Soviet Natural Gas and the Western European Energy Crisis**	76
	The great gas transaction — I	77
	The great gas transaction — II	81
	Alternatives to Soviet gas	88
	Conclusion	90
	Appendix: Soviet oil industry	91
6	**OPEC and Other Developing Countries**	96
	Background	96
	Natural gas and OPEC industrialisation	98
	Non-OPEC developing countries	100

CONTENTS

7 The Pacific Region and Canada — 105
   Natural gas in Australia — 109
   Australian Resource Rent Tax and the Gregory Thesis — 113
   Malaysia and New Zealand — 114
   Canada — 116
   Appendix: natural gas shipping — 118

8 The Western European Natural Gas Economy — 119
   Introductory survey — 119
   Trade, pricing and storage — 125
   Storage — 132
   United Kingdom, France and West Germany — 134
   Italy, Denmark and The Netherlands — 140
   Norway — 144
   Appendix: algebraic comment on security of supply in the Western European gas market — 149

9 Economic Theory and Natural Gas Pipelines — 153
   Depletion of gas fields — 153
   Natural gas pipelines: introduction — 158
   Pipeline economics: a neo-classical formulation — 163
   Average and marginal cost — 169
   Conclusion — 174

10 Energy and the Macroeconomy — 175
   The international economy and financial market disequilibria — 179
   Futures markets — 181
   Conclusion — 187
   Appendix: short-run oil pricing and the reserve/production ratio — 190

Bibliography — 194
Index — 198

Figures

1.1	Components of natural gas	5
3.1	Simple flow diagram of gas transmission process	42
3.2	Cost of transporting natural gas	47
3.3	A typical LNG export chain	52
3.4	Production/investment profiles	55
4.1	Price changes as deregulation takes place	74
5.1	Coordination and Nash-type games	86
8.1	LNG projects in the Atlantic Basin	127
8.2	Hypothetical demand curve for natural gas	128
9.1	Investment and production profile of a natural gas field	154
9.2	Average gas well production in the United States (1965–84)	156
9.3	Gas source, compressor stations, pipeline stretches and pressures in a simple natural gas transmission pipeline	159
9.4	Two production activities, A and B, for a pipeline	168
9.5	Cost curves given changes in V and Q	170
9.6	Marginal and average cost curves when volume is explicitly considered	173

Tables

1.1 Symbols used for gas units and conversion factors	7
1.2 Proved reserves (1985) and production (1984) of natural gas, in billions of cubic meters (Gcm)	10
1.3 Estimated consumption and percentage shares of natural gas (Gcm)	11
1.4 Natural gas exports in billions of cubic meters	14
2.1 World crude oil production (1,000 barrels/day); world oil reserves in billions of barrels (Gbbl); and the reserve/production ratio, in years	21
2.2 Coal production, exports, imports and reserves	24
2.3 Coal consumption (1985), and forecast consumption (1990, 1995 and 2000), coal trade (1985) and forecast trade (1990, 1995, 2000)	25
2.4 World trade in natural gas by importing countries (1984–5)	32
3.1 Operational LNG projects and projects under construction; proposed projects at 1 January 1984	50
4.1 Spot market prices in the United States (March 1985)	65
5.1 Forecast supply and demand of natural gas in the European Community for 1990 and 2000	84
5.2 Soviet natural gas exports to Western and Eastern Europe	85
5.3 Soviet exports of oil	91
7.1 LNG shipping (1986)	118
7.2 Oil tankers and combined carriers (1984, 1985)	118
8.1 Natural gas consumption and share of gas in six Western European Countries (1984)	120
8.2 Gas storage facilities in OECD countries, including facilities under construction in Mcm (1983)	133
8.3 Sources of French natural gas	137

For my colleagues and students in Sweden and Australia

Preface

This is the seventh volume in the Croom Helm Commodity Series. Other books which have been published in the series to date are *Uranium* by Marian Radetzki, *Tin* by William Robertson, *The Modern Plantation in the Third World* by Edgar Graham with Ingrid Floering, *Commodity Models for Forecasting and Policy Analysis* by Walter Labys and Peter Pollak, *The International Grain Trade* by Nicholas Butler and my own contribution, *International Commodity Control*. Volumes which are imminent include:

T MacBean and D T Nguyen — Commodity Policies: Problems and Prospects
B S Chimni — International Commodity Agreements: A Legal Study

Volumes covering a range of other subjects are also planned over the course of the next few years.

The aim of the Series is a general one: to advance the understanding of issues relating to the production and marketing of primary commodities. As a result, volumes in the Series deal with a range of subjects — from the examination of a fairly specialised commodity, such as uranium at one end of the spectrum, to a much broader subject, such as commodity price behaviour, at the other. Wherever possible, however, it is hoped that volumes in the Series will share a common form so that they may be useful for reference purposes. It is also intended that they should each be forward-looking rather than merely an historical account.

Contributors to the Series come from a variety of countries and backgrounds so that a variety of approaches have been adopted. In general, however, they are already established in their chosen fields and we are confident that this is being reflected in the quality of the Series as a whole.

The Series should appeal to anyone interested in commodity issues, whether their principal concern is with policy, marketing, trading or simply obtaining general background information. Both the publishers, Croom Helm, and myself, Fiona Gordon-Ashworth, as Series Editor, welcome any feedback that users of the Series may have.

Fiona Gordon-Ashworth
(Series Editor)

1

Introduction:
Geology, Units and Background

This is a book on applied energy economics. It is intended as a simple, comprehensive and up-to-date review of the natural gas markets. Unlike my books *The political economy of oil* and *The political economy of coal*, it concentrates almost exclusively on a single resource, conventional natural gas, although Chapter 2 provides a general energy outlook and the short final chapter considers some macroeconomic aspects of the 'energy crisis'. Here it seems useful to mention that at both the 1984 (Cambridge, England) and 1985 (Bonn, West Germany) meetings of the International Association of Energy Economists, it was widely maintained that during the remainder of this century, natural gas may have the most exciting future of all the energy resources. Certainly this should be the case if the large scale exploitation of the enormous natural gas supplies of the Middle East could begin.

In recent years a wide selection of books on natural gas has appeared. Without exception these studies were essentially bare of the analytical tools of mainstream economic theory. This is apparently what many of the readers of books on natural gas desire, and having examined much of the so-called econometric literature on natural gas, I understand their reluctance to confront mathematical symbols, and in particular to be forced to work through complicated algebraic materials in order to obtain results that could have been given in verbal form.

There is no econometric game-playing in this study, but the occasional supply and demand curve is used, as well as a small amount of algebra. However these materials have been located in such a way as not to interfere with the exposition, and the reader who is uninterested in this kind of presentation may ignore them.

On the other hand, one entire chapter is moderately technical.

INTRODUCTION: GEOLOGY, UNITS AND BACKGROUND

This is the penultimate chapter, and it mostly deals with natural gas pipelines. It is intended as an elementary introduction to a fairly difficult topic which, with certain exceptions, has been badly treated in the literature of economics.

FOREWORD

Natural gas has been called, and justifiably, the Prince of Hydrocarbons. It has also been labelled *Gaz Fatal*, because at one time natural gas was responsible for hundreds of deaths every year. These fatalities came about mainly through blowouts caused by expanding gas associated with oil, as well as gas explosions in households resulting from the deficient processing of natural gas. Problems of this type now appear to be extinct, and the main complaint associated with natural gas is that there is not enough of it within the World Outside the Centrally Planned Areas (WOCA).

Whenever and wherever gas became available, the speed at which it was adopted was phenomenal. The discovery of large amounts of gas at Groningen completely changed the energy picture in the Netherlands; and similar phenomena were observed in the United States and the Soviet Union. The Groningen gas field was discovered in the late 1950s, and by 1970 it supplied 45 per cent of all the primary energy consumed in The Netherlands. (Primary energy is energy obtained from the direct burning of coal, gas, oil, etc. as well as electricity having a hydro or nuclear origin. Electricity obtained from the burning of substances such as coal is a secondary energy source.) A large part of The Netherlands' welfare system was probably subsidised by this gas, and now that the Dutch gas fields are reaching an advanced stage of their depletion cycle, the Dutch welfare system — as it presently exists — may no longer be viable. In the United States, the natural gas share of primary energy consumption went from 15 per cent in 1945 to 32 per cent in 1965, although it began declining soon afterwards, and is now about 26 per cent. It is expected to fall to between 15 and 22 per cent in the next 20 years. Given the huge quantities of gas now becoming available by pipeline from the Soviet Union, the North Sea and Algeria, Western Europe may be on the threshold of a new energy dawn. Here it should be emphasised that because of its cleanliness — which is a quality it does not share with coal and oil — natural gas may be the ideal energy medium for central and northern Europe.

In the remainder of this chapter I provide general remarks on the chemistry of natural gas and its evolution as a prime energy medium. Then I explain the units in which natural gas is measured. Finally, I present a brief discussion of reserves and production. The main purpose of this chapter is to provide the reader with the vocabulary needed to follow the analysis of natural gas markets presented in this book, and elsewhere; and also to offer insights into the kinds of issues that are important for these markets.

GEOLOGY AND HISTORY

In general, natural gas is found in an environment similar to that in which crude oil originates, and on occasion it has been called gaseous petroleum. (Petroleum is oil plus gas.) The hydrocarbons of natural gas are, however, lighter and less complex than those of crude oil; and natural gas occasionally contains water and gases that are not hydrocarbons. Although many people believe that gas and oil are found in reservoirs, or huge underground caverns (and this terminology is in general use), they actually originate in water-coated pore spaces in rocks — for the most part sedimentary rock classified as organic shale. This shale originated as the remains of prehistoric plants and animals, and was 'cooked' into oil and gas by heat, the pressure of the earth acting over millions of years, and various chemical reactions.

Some hydrocarbon deposits contain oil but no gas, while others — where the cooking referred to above continued until the hydrocarbons were reduced from liquid oil to molecules of gas — contain gas but no oil. This latter category is called non-associated gas. A very common arrangement is the presence of gas and oil in the same deposit, and in this situation the gas is referred to as associated gas. Since non-associated gas may come into contact with water, but not oil, its production is discretionary: i.e. it can be shut in (left undeveloped) until the market warrants its exploitation. On the other hand, associated gas is not discretionary because it becomes available whenever the associated crude oil is produced, and thus if it is not piped away to be sold it must be reinjected in order to maintain the pressure of the reservoir, or flared (burned up in the air). Associated gas can be found dissolved in oil, or sitting in a gas cap on top of the oil, or both. If initially there is no gas cap, the depletion of the well can create one by lowering the pressure inside the reservoir, and thus releasing some of the dissolved gas in the form of a liquid. This dissolved gas contains for the most part light components: e.g. methane, ethane, and liquefied petroleum gas (LPG). If the gas is not — or cannot be — tapped when oil is being produced, then it may be tapped after the oil well has been closed.

In 1983 about 106 billion cubic meters (= 106 Gcm) of gas was flared, which was about six per cent of gross production. Most of this flaring took place in the Middle East and Africa, where the largest proportion of associated gas is located. Another major but little known source of losses is due to treatment and liquefaction: in 1983 this came to 64 Gcm. In the same year reinjection was about 8 per cent of gross production, or 152 Gcm; here it should be noted that reinjection often, but not always, permits producer countries to delay their decision on the use of associated gases, since up to 80 per cent of this reinjected gas can be recovered after oil production has ceased — depending on the nature of the reservoir. Four countries accounted for 75 per cent of total reinjection: Algeria, the United States, Venezuela and Canada. It is believed that gas can be found at much greater depths than oil, and new drilling technologies are being developed that will permit probes at between 30,000 and 50,000 feet. Once exploration is fully underway at these depths, it is expected that world reserves of gas will show a sizable increase. It is also thought that a great deal of unconventional natural gas can be found in coal seams, tight sands, deep basins, beneath hydrate layers that form under permafrost and the floor of the seas, and offshore in faults where salt water saturated with dissolved gas collects.

Conventional non-associated natural gas from a well consists mainly of methane — on the average about 85 per cent; heavier hydrocarbons known as natural gas liquids (such as ethane, propane, butane, pentane, and some heavier fractions), water, carbon-dioxide, nitrogen and some other non-hydrocarbons. On the other hand, associated gases may contain between 20 and 50 per cent of gas streams other than methane. Methane can be thought of as pure natural gas, while the other gas streams are used as petrochemical feedstocks, fuels and inputs for refinery processes. Figure 1.1 shows the components of natural gas, ex-well.

Before dry natural gas can be distributed to consumers, undesirable components must be removed and a uniform quality obtained by decreasing the share of hydrocarbons. Natural gas liquids are separated out, and where a market exists the most valuable components — butane and propane (i.e. LPG) — are sold, usually under the name *gasol* or bottled gas. In those countries where the

Figure 1.1: Components of natural gas

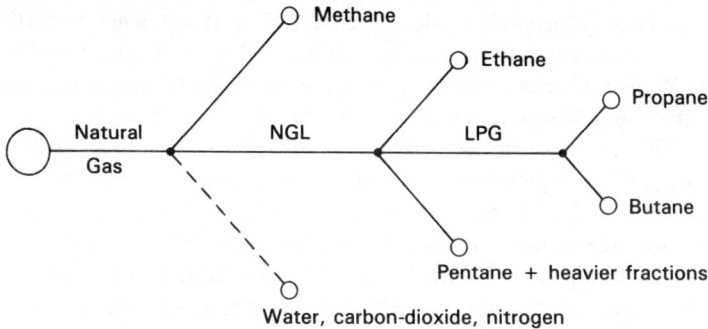

	Liquefied at	Definitions
Butane (normal)	−0.5°C	NGL: natural gas liquids
Butane (Iso)	−12.0°C	LPG: liquefied petroleum gas
Propane	−42.0°C	LNG: liquefied natural gas
Ethane	−88.0°C	SNG: synthetic (substitute) natural
Methane	−161.0°C	gas

Notes: natural gas = methane + NGL + (water, nitrogen, CO_2).
NGL = ethane + LPG + (pentane and heavier fractions).
LPG = propane + butane + mixtures of propane and butane.

availability of natural gas is greater than the absorption capacity of local markets, and where for one reason or another it is not possible to transport the gas to foreign markets in its original form, the processing of natural gas into gas liquids is a valuable economic activity.

Before leaving this section, something must be said about the history of natural gas. It has been claimed that natural gas was known and used in China before the birth of Christ. The word gas was probably invented by the Dutch chemist Jan van Helmont in 1609, although the expression natural gas was coined in 1795 by the Italian Lazzaro Spallanzani. The term methane is attributed to the German chemist August Wilhelm von Hofmann, while the Swedish scientist Jöns Jacob Berzelius first presented the chemical formula for methane (CH_4).

Initially the gas industry was based on town gas (or manufactured gas), which is not natural gas but a gas manufactured by carbonising coal. This fuel was first introduced to the larger cities of the world in 1812 (in England), and is still used in the German Ruhr. The

United States first used town gas in 1816, and five years later the first use of natural gas was recorded at Fredonia, New York. The first long distance all-welded pipeline (of 14 to 18 inches and 217 miles long) was put into operation between Louisiana and Texas in 1925, and this can probably be taken as the beginning of the modern natural gas industry.

The most 'natural' way of moving gas is via pipelines, at least up to distances of about 8,000 kilometers. Another option is to cool the gas to $-162°C$ Centigrade by a cryogenic cooling process. This reduces the volume of the gas by a factor of 600, and from the liquefaction plant the liquefied natural gas (LNG) is transferred to a cryogenic carrier and transported to the consuming country, where it is regasified. Some of these ships can carry an amount of gas equivalent to 458,000 barrels of oil. It is sometimes claimed that LNG trade is more flexible than piped natural gas because LNG carriers can be rerouted, while pipelines are fixed. In considering the long term contracts under which most gas is sold, however, I am not sure that this argument is correct. One thing that must be emphasised is that transporting gas in any form is more expensive than transporting an equivalent amount of oil (measured in heating or calorific value) by a comparable carrier.

Natural gas production takes place both on and offshore. Offshore production has grown by 14 per cent annually over the last decade to reach about 20 per cent of all commercial production. The United States accounts for almost 50 per cent of world offshore production, and one fifth of known United States natural gas reserves are offshore. Western Europe now accounts for 30 per cent of world offshore production, mostly in the Norwegian, British and Danish North Sea. In the long run it is believed that the Arctic seas offshore from Norway and the Soviet Union will be of extreme importance to world natural gas.

UNITS AND CONVERSION FACTORS

Next we come to units and conversion factors, and since the units that are associated with gas are much more unfamiliar than those associated with oil and coal, the discussion below will be more thorough than is usually the case. The designation one metric ton (=1 tonne = lt) equals 2,205 pounds (lbs). In everyday life the usual ton is the short ton, or simply ton, which equals 2,000 pounds, and thus lt = 1.1023 tons. Finally, there is a long ton, which equals

Table 1.1: Symbols used for gas units and conversion factors

Prefix	Symbol	Power	Meaning	Example
kilo	k	10^3	thousand	kW (kilowatt)
mega	M	10^6	million	MW (megawatt)
giga	G	10^9	billion	GW (gigawatt)
tera	T	10^{12}	trillion	TW (terawatt)
peta	P	10^{15}	thousand-million	PW (petawatt)
exa	E	10^{18}	million-million	EW (exawatt)

2,240 pounds. In case the reader prefers working with kilograms, then it should be observed that 1 kilogram = 2.2 pounds (and 1 inch = 2.54 centimeters).

In the matter of energy we are usually interested in heat equivalents. A common unit is the British thermal unit (or Btu), which is the amount of heat required to increase the temperature of one pound of water by 1 degree Fahrenheit, where one pound is approximately equal to one pint. One pound of coal has an energy content of between 10,200 and 14,600 Btu, while crude oil has an energy content of between 18,300 and 19,500 Btu/pound. One thousand cubic feet (= 1000 cf = 1 kcf) of natural gas has an energy content of 1,000,000 Btu on average. In scientific work, and in certain countries, joules are preferred to Btus as a unit of heat energy. The transformation here is straightforward: 100,000 Btu = 1.055×10^8 joules, or 1 Btu = 1.055×10^3 joules.

It takes approximately 7.33 barrels to 'enclose' lt of crude oil, where the exact figure runs from 6.98 barrels/tonne (= 6.98 bbl/t) to 7.73 bbl/t. Using the conversion factors given in the above paragraph, lt of crude oil contains between 40,351,500 Btu and 42,997,500 Btu. An unweighted average thus indicates that 1 barrel (= 1 Bbl) of oil = 5,685,470 Btu, although a handy transformation when the heterogenity of oils is taken into consideration is 1 bbl oil = 5,800,000 Btu. Here it can also be noted that the most popular unit for measuring both the production and consumption of oil is millions of barrels per day (Mbbl/d). Furthermore, barrels/day can be turned into Mt/year by multiplying by 50. For example, 20 Mbbl/d = 20 × 50 = 1,000 Mt/year. The truth of this relationship can immediately be seen from a simple dimensional analysis:

$$\left(\frac{\text{barrels}}{\text{day}}\right) \times 365 \left(\frac{\text{days}}{\text{year}}\right) \times \frac{1}{7.33} \left(\frac{\text{tonnes}}{\text{barrel}}\right) \approx 50 \left(\frac{\text{tonnes}}{\text{year}}\right)$$

At this point we turn to natural gas. It was already mentioned above that 1,000 cubic feet of natural gas has an approximate energy content of 1,000,000 Btu, and the first thing that can be noted here is that 1,000 cubic feet is equal to 28.3 cubic meters, since 1 cubic meter = 35.3147 cubic feet. (To be exact, 1 cubic foot of natural gas has an average heating value of 1035 Btu, but 1,000 Btu is almost always used.) In terms of heating values, 1,000 cubic feet of natural gas is the equivalent of 0.179 barrels of crude oil, which means that 1,000 cubic meters of natural gas is approximately the equivalent of $(1,000/28.3) \times 0.179 = 6.3$ barrels of crude oil, on average. The price of natural gas is often quoted in terms of MBtu; for example, the average price of the gas that was to be originally supplied by the Soviet Union to Western Europe on contracts associated with the Soviet-Western Europe pipeline was 4.65 dollars/MBtu. It is instructive to compare this price to the official price of oil at that time, which averaged about 32 dollars/barrel. Using the above conversion units, with an average barrel of oil containing 5,800,000 Btu, this gives an equivalent gas price of 27 dollars/barrel. The contracts on which this gas was bought contained some index clauses, and thus the price of this gas has recently fallen below 4 dollars/MBtu, or 23 dollars/bbl of oil equivalent. This latter figure can be compared with the latest official price of oil, which was 28 dollars/bbl.

Another important unit of measurement where natural gas is concerned is the calorie, or the kilocarie (= 1,000 calories). The transformation here is also simple: 1 calorie = 4.184 joules; and 1 Btu = 252 calories. In Australia the *therm* (= 100,000 Btu) is occasionally used, and in the United States the quad (= 10^{15} Btu).

I would like to complete this discussion by saying something about the concept of an equivalent unit of coal, since as implied above the price of one energy medium (e.g. gas) will often be compared to the price of another (e.g. coal or oil); but even the same weight or volume of a particular energy medium can have different heating values because of the essential lack of homogeneity within — as well as between — energy media. A ton of coal equivalent (tce) is defined as a metric ton (= tonne) of coal with a heating value of 12,600 Btu per pound. Consequently, since the heat energy content of e.g. bituminous coals ranges from 10,200 Btu to 14,600 Btu, more than a tonne of some bituminous coals is required to produce the heating value of 1 tce. Consider also that in 1977 world coal production was 3,400 million metric tons of raw coal, which was 2,500 million metric tons of coal equivalent (2,500 Mtce), or 33 million barrels of oil per day (= 33 Mbbl/d). This last figure is obtained from the following conversion scheme.

One tonne of coal equivalent (= 1 tce) is equal to $1 \times 2{,}205 \times 12{,}600 = 27{,}783{,}000$ Btu. 1 bbl oil = 5,800,000 Btu, so 1 tce = 4.8 bbl oil. In addition, 1 Mbbl/d of oil = 365 Mbbl/year (= 365 Mbbl/y), which by the above conversion is 365/4.8 = 76 Mcte/y. Thus 2,500 Mtce/y = 2,500/76 = 33 Mbbl/d. Let us now go to natural gas. The production of natural gas in 1984 was 1,570 Gcm, which is equal to $1{,}570 \times 35.317 = 55{,}444$ Gcf. Since 1,000 cubic feet gives 1,037, 000 Btu (using the exact heat content instead of the approximate value of 1,000,000 Btu), we get as the Btu content of 55,444 Gcf of gas:

$$\frac{55{,}444 \times 10^9 \times 1.037 \times 10^6}{10^3} = 57.38 \times 10^{15} \text{ Btu}$$

But, as above, 1 bbl oil = $5{,}800{,}000 = 5.8 \times 10^6$ Btu on the average, and so 57.38×10^{15} Btu is equal to:

$$\frac{57.38 \times 10^{15}}{5.8 \times 10^6} = 9.8 \times 10^9 \text{ bbl/year}$$

Dividing by 365 gives 26.8 Mbbl/day.

RESERVES, PRODUCTION, USE AND TRADE IN NATURAL GAS

The world reserves of natural gas were, as of 1 January 1985, over 90,000 billion cubic meters (= 90,000 Gcm = 3,178 Tcf). They have been growing at an average annual rate of 5 per cent since 1980, and the current reserve/production ratio is almost 59 years. As made clear in Table 1.2, the Soviet Union has the largest gas reserves, with about 40 per cent of the world total; and in addition new reserves are probably being found in Russia at a more rapid rate than elswhere. Iran is in second place with about 11 per cent of world gas reserves, and altogether the Middle East has 25 per cent of world reserves. More than one expert has claimed that the beginning of large scale exploitation of Middle Eastern reserves will be the energy event of this century. There has also been a substantial adjusting up of reserve estimates in Western Europe, particularly in Norway.

In terms of oil equivalencies, 90,000 Gcm of gas is the equivalent of 562 billion barrels of oil, which is at least 80 per cent of world

Table 1.2: Proved reserves (1985) and production (1984) of natural gas, in billions of cubic meters (Gcm)

	Proved reserves, Gcm (as of 1 January)		Per cent of reserves		Production (Gcm)	
	1984	1985	1985	1985*	1983	1984
North America						
Canada	2,613	2,660	2.9	(78)	71.34	78.19
United States	5,645	5,670	5.9	(475)	450.20	487.50
Caribbean and Latin America						
Mexico	2,180	2,172	2.3	(32)	31.11	29.35
Trinidad	430	350	0.4		3.55	5.43
Venezuela	1,545	1,650	1.7		16.25	17.26
Others	117	120				
Other Latin America						
Argentina	680	673	0.7		12.55	13.98
Others	479	479				
Middle East						
Abu Dhabi	2,700	2,650	2.8		6.50	7.66
Iran	11,380	13,550	14.1	(13.7)	8.90	13.50
Iraq	821	760	0.8		0.65	0.59
Kuwait	1,030	985	1.0		4.50	4.10
Qatar	3,400	4,280	4.4		4.73	5.93
Saudi Arabia	2,188	2,350	2.4	(13.7)	5.53	7.15
Others	917	1,034				
Africa						
Algeria	3,155	3,087	3.2	(31)	35.59	38.53
Libya	555	570	0.6		4.05	4.60
Nigeria	1,370	1,330	1.4		2.00	2.75
Others	662	660				
Western Europe						
Netherlands	1,927	1,890	2.0	(74)	72.98	75.14
Norway	2,039	2,236	2.3	(26)	24.42	27.28
United Kingdom	712	725	0.8	(41)	39.53	40.16
Others	662	660				
Far East						
Australia	945	1,482	1.5	(17)	14.16	11.42
Thailand	240	210	0.2		1.61	2.40
China (Mainland)	800	850	0.9		19.80	18.00
Indonesia	1,000	1,100	1.1	(33)	20.83	30.05
Malaysia	1,400	1,390	1.4		3.70	9.20
Pakistan	510	600	0.6		9.72	10.04
Others	1,495	1,611				
Eastern Europe						
Soviet Union	26,000	37,500	39.0	(650)	535.95	587.00
Others	638	666				
World total:	90,325	96,197	100.0			1686.40

* Estimated
Source: OPEC bulletin (various issues); *The Petroleum Economist* (August 1985)

proved reserves of crude oil. Thus before this century ends, gas is capable of becoming a much more important source of energy than it already is, although for this actually to happen there would have to be an enormous amount of investment in gas distribution systems, and probably the beginning of the widely talked about methanol economy. It can also be mentioned here that in addition to the approximately 27 Mbbl/d of gas used in 1984 (measured in oil equivalency), about 60 Mbbl/d of oil was produced, 35 Mbbl/d of coal, 10 Mbbl/d of hydro energy, and 4 Mbbl/d of nuclear energy.

In 1984 the production of natural gas increased for the first time since 1979, and together the United States and the Soviet Union accounted for over 75 per cent of the world increase in production. Sixty-five per cent of global gas consumption is concentrated in these two countries, and in fact 85 per cent can be localised to just ten countries. Thus, in contrast to oil, most of the gas that is consumed is produced locally. The largest single increase in consumption was in the Soviet Union, where a 45 Gcm increment in consumption made that country the world's largest consumer of natural gas. In the United States there was a rise in consumption of 4 per cent, which reversed the declining trend of the previous five years. As Table 1.3 indicates, consumption was also up in Western Europe and in Japan,

Table 1.3: Estimated consumption and percentage shares of natural gas (Gcm)

	1984	%	1985	%
United States	490	31	475	29
Canada and Mexico	77	5	80	5
Total North America	567	36	555	34
Soviet Union	510	33	560	34
Eastern Europe and China	100	6	105	6
Total USSR, East Europe, China	610	39	665	40
Japan	38	2	41	3
Australasia (and others)	44	3	52	3
Total Japan, etc.	82	5	93	6
United Kingdom	51	3	54	3
West Germany	49	3	49	3
Netherlands	36	2	38	2
Italy	31	2	38	2
France	27	2	28	2
Others	21	2	22	1
Total Western Europe	215	14	223	13
Rest of world	96	6	109	7
Total world	1,570	100	1,645	100

Note: All percentages are rounded
Source: Shell briefing reports (1983, 1984, 1985, 1986)

where in the latter country the input of gas into power generation increased by approximately 40 per cent, mostly at the expense of oil.

The markets for natural gas can be divided into three main categories: electricity generation, industry, and domestic and commercial. In the United States and Western Europe the latter of these is regarded as a prime market, and it accounts for between 40 and 50 per cent of all gas consumed. But in the Soviet Union well over half of natural gas consumption takes place in the industrial sector, while in Japan electricity generation accounts for almost 70 per cent of the gas consumed, and about 14 per cent of all the electricity produced employs natural gas as the primary input. The strange thing about this is that approximately 45 per cent of Japanese households are connected to the gas grid; however pollution regulations have been tightened considerably, and power stations close to cities are under considerable pressure to burn gas instead of coal or oil.

The United States can be regarded as a highly mature gas market, where consumption probably peaked in 1973 (at 609 Gcm). But even so natural gas is still the second most important fuel after oil, and accounts for slightly more than 25 per cent of primary energy consumption. The picture in Western Europe is considerably different. Reserves are much smaller than in North America, and market penetration on a wide scale did not commence until discovery of the huge gas deposits at Groningen, The Netherlands. Later, major supplies were discovered in the British and Norwegian sectors of the North Sea, and Norway was able to join The Netherlands as a significant exporter of gas. On the whole, natural gas consumption in Western Europe has been able to grow at about 15 per cent/year since 1960, and at present natural gas constitutes 14 to 15 per cent of Europe's primary energy. In 1982 over 85 per cent of Western Europe's total gas supply came from indigenous resources, but this figure should fall rapidly as more gas is imported by pipeline from North Africa and the Soviet Union, and also in the form of LNG from North Africa.

The Soviet Union is becoming one of the world's great exporters of natural gas, and after an average yearly production growth of 14.5 per cent since 1975, overtook United States production in 1982 (of 488 Gcm = 18.3 Tcf) to become the world's largest producer. The USSR has imported some gas from Iran and Afghanistan, but for the most part that country produces all the gas it consumes. In recent years a number of giant gas fields have been found in Western Siberia, and larger volumes will undoubtedly be exported when markets become available. Consumption of natural gas will also rise rapidly. The Soviet Union is one of the most energy intensive countries in the world, and natural gas will undoubtedly have to replace oil in many of its uses if the production of oil declines, which is quite possible. Similarly, the Soviet Union is attempting to convince Eastern Europe to use more Soviet gas and less Soviet oil. At present the countries of Eastern Europe, to include the Soviet Union, constitute the major world gas-consuming region, accounting for almost 37 per cent of the total amount of global gas consumption. Japan has no significant indigenous gas resources, and when LNG imports started in 1969, natural gas contributed less than one per cent to total primary energy consumption. At the present time natural gas accounts for slightly more than seven per cent of the primary energy used in Japan. The policy of Japan is to import energy resources from as many producing countries and regions as possible, and as things now stand, Japan is the major importing country in the world energy market. This observation applies to crude oil and steam coal as well as natural gas.

The three main components of natural gas trade are production, transportation and distribution. The first of these includes processing and treatment to remove undesirable constituents, as well as obtaining a uniform quality that is characterised by the absence of heavier hydrocarbons.

The transportation phase involves moving gas from producing to consuming areas by large diameter, high pressure pipelines (which also serve to move gas from deep-water offshore fields to onshore installations), or by ship in the form of LNG. The most modern LNG vessels can carry the equivalent of nearly 80 million cubic meters of gaseous natural gas per voyage, which is the equivalent of 458,440 barrels of oil. Finally, local distribution also involves pipelines, but in this case medium and low pressure pipeline grids to final (or end) consumers. For important insights into this topic see Percebois (1986).

The international trade in natural gas originated slightly over 90 years ago with the sale of pipeline gas by Canada to the United States. Recent major pipeline projects include the supply of Dutch and Norwegian gas to various Western European countries; Soviet gas to Eastern and Western Europe; Mexican gas to the United States; and, since the last part of 1983, the supply of Algerian natural gas to Italy via the trans-Mediterranean pipeline. (This pipeline was completed two years earlier, but the Algerians refused to provide gas because they wanted higher prices for their gas than originally agreed upon.) Later projects were the shipment of LNG from North Africa to Europe and the United States; shipments by Brunei, Alaska, Abu Dhabi, Indonesia and Malaysia to Japan, as well as deals that are now completed for shipment by Australia and Thailand to Japan. It seems likely that in the distant future Iran will figure as a major supplier of LNG, and if this happens Iran could eventually dominate the supplier side of the LNG market. Table 1.4 presents a simple breakdown of the international natural gas market by exporters.

Table 1.4: Natural gas exports in billions of cubic meters

	1984	%	1985	%
Exports by pipeline from:				
Soviet Union	62	31	65	31
Netherlands	32	16	33	15
Norway	26	13	26	12
Canada	21	10	24	11
Others	13	6	14	7
Total pipeline exports	154	76	162	76
Exports as LNG from:				
Indonesia	19	9	20	9
Algeria	12	6	13	6
Brunei	7	4	7	3
Malaysia	5	2	6	3
Others	6	3	6	3
Total LNG exports	49	24	52	24
Total exports	203	100	214	100

Source: Shell briefing reports (1984, 1985); OPEC bulletin (various issues); IEA documents.

At present the annual international trade in natural gas is only a small part of world consumption but major increases are expected. Canadian and Mexican exports should replace an appreciable amount of domestic United States production, for example. The future status of Japan as an importer of LNG is uncertain, but while it is likely that less LNG will be used in the power generating sector of that country, more gas will probably be used in households. In any event, the exports of LNG to the Far East should approximately double by 1990, and if new markets are opened up in Taiwan and South Korea, as expected, the increase could be much larger. At present, gas being traded in LNG form amounts to almost one quarter of the total amount traded (with the remainder being traded by pipeline), but the ratio of LNG to pipeline gas is on an upward trend, and this tendency will almost certainly continue.

I close this chapter with a few general remarks on the trade in natural gas. There is little question that this trade will expand, although perhaps not at the rate expected a decade ago. Furthermore, these expectations led to an over-investment in natural gas-producing facilities, and of late to a surplus of gas. The average price of internationally traded natural gas has probably dropped by 35 per cent since 1981, thereby threatening the profitability of a number of existing installations, and causing the cancellation of a number of projects that were once considered almost certain to come into existence. With most gas prices under 4 dollars/MBtu in February 1987, it seems odd that any new projects at all should be underway, since only a few years ago most producers in the World Outside the Centrally Planned Areas (WOCA) insisted that new projects would require gas prices of 6 to 7 dollars/MBtu if they were to be viable.

For years the producers of natural gas strove to convince consumers that the price of their product should be brought into line with crude oil. Now this is taking place, but in a manner considered unfavourable by producers: crude oil prices are falling, and existing indexing formulae have led to the price of gas falling at an even more rapid rate. The present situation has also brought about a virtual collapse of the traditional three-tiered price structure for internationally traded gas, where the highest price was paid by Japan, the lowest by the United States, and Western Europe was in between. At the present time the market price of traded gas is almost uniform, and fairly close to the crude oil equivalent at the point of delivery.

It has been claimed that in the absence of export markets, more gas exporting countries, and potential exporters, will attempt to increase the domestic consumption of gas. This may or may not be true, since revenue from such things as the export of gas is often needed to finance investments required to use gas in power generation and petrochemical activities. On the other hand, the manufacture of such things as fertiliser and other relatively simple petrochemical products from gas often makes good economic sense. Regional arrangements may also be appropriate in certain situations. For example, electricity can be generated in a gas-rich country and sold in countries several thousands of miles away.

2

Energy Trends

The purpose of this chapter is to examine the world energy scenario, and in particular that part relating to oil, coal and natural gas.

Nineteen eighty-four was the first year since 1979 that there was a marked upturn in the consumption of energy resources. Total world production of crude oil and natural gas liquids increased by over two per cent to 58.2 million barrels per day (= 58.2 Mbbl/d). Natural gas production was 1,570 billion cubic meters (= 1,570 Gcm), which was an increase of seven per cent over 1983, and the international trade in natural gas increased by almost nine per cent. World hard coal production was slightly over three billion tonnes ($\dot{=}$ 3 Gt), which was an increase of three per cent over 1983. The international coal trade exceeded 300 million tonnes (= 300 Mt), which was an increase of 19 per cent on the previous year.

Where oil was concerned the question was, as always, when would the recovery in demand take place? and is it possible to make a meaningful statement about the future of the oil price? This matter will be approached in the first section of this chapter, where the principal concern will be the long run situation — which means focussing on oil reserves rather than the production and consumption of oil. An important observation concerning the supply of oil in the long run will be made, in algebraic form, in the appendix to this chapter. I also want to call specific attention to the latter part of this chapter, which continues on an elementary plane the discussion of natural gas begun in the previous chapter.

OUTLOOK FOR OIL

Every day sees an increase in the concentration of oil reserves (known recoverable oil in the ground) to a relatively few producing countries, mostly in the Middle East. OPEC already controls three-fourths of oil reserves outside the USSR, China, and East Europe — or 448 of 593 billion barrels — and by the end of the century may have as much as four-fifths. This has belatedly become an accepted fact but the importance of two related phenomena seem to have been overlooked. First, geologists associated with such highly regarded organisations as the United States Geological Survey and the Library of Congress's Congressional Research Service have adjusted downwards their estimates of the total amount of oil that will be eventually recovered. Furthermore, it is now contended that only a small possibility exists that a massive new oil province will be discovered, at least where it will do the industrial countries outside the centrally planned areas any good. This represents a complete turnabout from the common belief of a decade ago that the discovery of at least one, and possibly two, supergiant oil fields were virtual certainties. It should also be appreciated that a large slice of the world's oil expertise considers it likely that if, or when, very large additions to reserves take place, they will be located in the Middle East.

Second, despite the most intensive exploration efforts in history, reserves are falling in the United States, Western Europe and the USSR; and with the exception of the Middle East, at least half the increase in world oil reserves listed since 1980 has been due to revisions of previous estimates of the amount of oil found in fields that were already being exploited. For technical reasons, the form in which reserve additions take place is extremely important, and if the goal of reserve replacement is to be realised, then large new fields will have to be found. Since exploration expenditures are now declining from the heights reached in 1980–1, and exploration results almost everywhere have been mediocre during the last decade, current output levels in the above countries cannot possibly be maintained in the longer term unless there is a drastic improvement in discovery rates. Consider, for example, North America, where reserves rose marginally in 1984, although they have fallen over the past three years by 7.3 per cent, and now account for under five per cent of the world total.

Just now world oil production is edging up, with an increasing diversity of sources of supply: in 1973 the ten largest producing countries accounted for 82 per cent of the world's output, while in 1984 this proportion had fallen to 75 per cent. As far as I am concerned, there is nothing special about the present pattern of world production, with the possible exception of the small decline that has taken place in the Soviet Union, which some observers see as the prelude to a very large decline. It is claimed however that the problem here is production rather than reserves, and the matter can be rectified by more investment and some changes in high level personnel. This may be true, but at the same time unless there are some very large discoveries in the Soviet Union in the fairly near future, production will have to decline. The same is true for some other large producing areas: the British North Sea, the United States and several OPEC countries. As yet it is impossible to say what will happen to oil exports from the Soviet Union to the capitalist world but it cannot be excluded that these might also come under a palpable downward pressure.

Where the outlook for reserves is concerned, it is useful to understand the way in which reserves influence production. In the United States proved oil reserves rose over the 1950s and 1960s to the all time high of 46.7 billion barrels in 1970. These have now dropped to 34.5 billion barrels, despite the greatest drilling boom in United States history. (More than 38,000 wells drilled in 1981 for a cost of almost 12 billion dollars, which represents a four-fold increase on the drilling expenses of 1971.) The domestic reserves/production ratio in the United States is presently about nine, and it is expected to reach eight by 1995. According to Tony Scanlan of the British Institute of Energy Economics, if exploration results in Alaska do not improve, the United States oil scene will be dominated by falling production in the lower 48, and the reserve/production ratio could fall as low as five or six. When this ratio is lower than ten, a real danger exists that the reserve base is being destroyed, and so I do not expect this ratio to reach five under any circumstances, short of war. Instead production will probably be decreased at a more rapid rate than anticipated by Scanlan. Similarly, British Petroleum's Magnus Field has been called the last of the North Sea dinosaurs, and in that part of the world oil companies will have to go after oil that, up to now, has been too costly to pump. This presages a steep drop in production sometime in the 1990s, despite the recent adjusting up of reserves that followed a Monte Carlo computer exercise somewhere onshore in the United Kingdom.

In Latin America the increase in reserves in 1984 was half that of the previous year, or 3.2 Gbbl, with the principal increase taking place in Venezuela — an OPEC member whose total reserves are now about twice those of the United Kingdom. An important new find also took place in Colombia; but on the other hand the United States Energy Administration claims that the Mexican reserves are about 18 billion barrels less than the 48.6 Gbbl claimed by the authorities of that country. Similarly, no significant discoveries were made in Africa in 1984, and on that continent reserves declined by 1.9 Gbbl.

In Asia reserves have been falling since 1982, although significant discoveries were made in 1984 in Malaysia, Indonesia, Pakistan and Australia. The source of the difficulty here is the disappointing prospecting and drilling picture in the offshore areas of China, where a potential once compared with the North Sea has proved to be considerably less. In the case of Western Europe, reserves rose by 1.4 Gbbl, principally in the North Sea, but very few observers really expect additions to reserves in that area to compensate for depletion for very many more years.

Reserves have shown a large increase in the OPEC countries of the Middle East, and this has taken place despite stagnation in the amount of drilling: in 1984 only 83 drilling rigs were in use in the Middle East, which was 60 per cent of the number operating in Western Europe. In 1984 OPEC added the oil equivalent of a present day Iran, or 50 Gbbl, which brought the organisation's share of world reserves to 68 per cent, as compared to 65 per cent in 1983. The overwhelming part of these new reserves belongs to the Middle Eastern members of OPEC. The reserve/production (R/P) ratio of OPEC is now 75 as compared to 17 years outside OPEC; and the R/P ratio of the Middle Eastern OPEC countries is 96 years. Here it seems appropriate to note that one of the major reasons for the discrepancy in R/P ratios between OPEC and the non-OPEC countries is the large amount of production in the latter as compared to OPEC. These production rates, of which some observers are so proud, almost guarantee that production outside OPEC must fall before the end of this century if the amount of reserves being found in these countries is not significantly larger than what is now being located. Let me sum up this discussion by saying that world reserves increased in 1984 by almost 30 Gbbl on a net basis, as compared to a slight decline in 1983; but all except a miniscule portion of this increase was accounted for by the 29 Gbbl net increase in the Middle East.

This situation is shown in Table 2.1, together with information about production and the R/P ratio for 1984 and 1985; however before presenting this table, something will be said about the often-cited contention that no realistic conclusions are possible about ultimate reserves of oil until as many boreholes have been drilled in Africa, South America and on other continents as have been drilled in the United States. As it happens, the non-uniform distribution of minerals in the crust of the earth is a geological law that no legitimate geologist expects to see refuted, and what it means in this context is that there are large areas of the world that possess little oil. Moreover, I consider the following counter-example instructive. Suppose Russia had not sold the United States Alaska; and Mexico had won the war between Mexico and the United States, and annexed the rest of Texas and also part of Southern California (which some Mexicans still regard as Mexican territory). This would have left the United States with an immense amount of land, and employing the logic that discoveries are directly proportional to the number of exploratory wells, it could have been argued at that time that the 'lower' 47 (= the 'lower' 48 minus Texas) should be rich in oil, particularly since it was flanked by two rich oil provinces, and there are only a few areas in the United States where some oil has not been found. However, tens of thousands of boreholes later we know that this is not the case. Without Alaska, Texas and perhaps also that part of California that once belonged to Mexico, the United States would not have been able to enjoy the huge benefits that derived from enormous oil wealth, despite the fact that the remainder of the United States is considerably larger than these three areas combined.

The reserve/production ratios in Table 2.1, as well as their likely development, indicate that unless the demand for crude oil collapses, the oil market should be back in balance before the middle of the 1990s, when the price will assume a definitive upward trend, although it is not likely that we will see any price explosions of the 1973–4 and 1979–80 varieties.

Something that has not been taken up above is OPEC's own demand for energy resources. But as it happens this will be growing as OPEC countries increase their production of petrochemicals and their standard of living increases. Moreover, OPEC's consumption of energy could, conceivably, begin to accelerate just as the demand for OPEC's oil by energy importers in the industrial world is turning upwards. Natural gas will supply a large part of the inputs for OPEC petrochemicals, which will be made clear later in this book; but this is not so significant as the fact that when petrochemical revenues increase, revenues from the sale of crude oil become less important. As things now stand, the world's 20 largest chemical concerns are in North America or Western Europe; but by the year 2000 at least

Table 2.1: World crude oil production (1,000 barrels/day); world oil reserves in billions of barrels (Gbbl); and the reserve/production ratio, in years

Canada	Production	Reserves		Reserve/production ratio	
	1985	1984	1985	1984	1985
Canada	1,700	6.7	7.1	13	14
United States	10,530	27.3	27.3	9	9
Total North America	12,230	34.0	34.4	9	9
Argentina	460	2.4	2.3	14	13
Brazil	580	1.8	2.0	16	12
Mexico	3,010	48.0	48.6	49	49
Venezuela	1,740	24.9	25.8	38	41
Other Latin America	950	4.5	4.6	—	—
Total Latin America	6,740	81.7	83.3	37	37
Iran	2,250	51.0	48.5	54	61
Iraq	1,430	43.0	44.5	130	100
Kuwait	1,070	63.9	90.0	192	266
Oman	500	2.8	3.5	20	24
Saudi Arabia	3,730	166.0	169.0	93	105
United Arab Emirates	1,350	34.0	34.0	—	—
Other Middle East	620	9.4	8.9	—	—
Total Middle East	10,950	370.1	398.4	87	96
Algeria	1,010	9.2	9.0	37	41
Egypt	910	3.5	3.2	14	12
Libya	1,100	21.3	21.1	57	53
Nigeria	1,500	16.6	16.7	37	32
Other African	930	6.3	5.0	—	—
Total Africa	5,450	56.9	55.0	34	33
United Kingdom	2,700	13.2	13.6	16	15
Norway	850	7.7	8.3	35	33
Other Western Europe	540	2.2	2.2	—	—
Total Western Europe	4,090	22.9	22.9	20	19
India	600	3.5	3.5	25	18
Indonesia	1,350	9.1	8.7	19	18
Malaysia	440	3.0	3.0	22	18
Australia	650	1.6	1.4	11	8
Other Asia Pacific	340	1.8	1.9	—	—
Total Asia Pacific	3,380	19.0	18.5	19	17
China	2,490	19.1	19.1	25	23
Soviet Union	12,000	63.0	63.0	14	14
Other Centrally Planned	420	2.5	2.0	—	—
Total Centrally Planned	14,910	84.6	84.1	16	15
Total non Centrally Planned	42,840	584.7	614.6	—	—
Total world	57,750	669.3	698.7	34	35
Total OPEC	17,300	448.0	476.4	70	75
Total non-OPEC	—	—	—	18	17

Note: All figures are rounded. Natural gas liquids are not included.
Source: OPEC bulletin (various issues); OPEC annual report (1983, 1984, 1985); Quarterly energy review (Economist Intelligence Unit, 1985, various issues); International Energy Agency annual report (1983, 1984).

ten of these will be in other parts of the world: Japan, the Soviet Union, Saudi Arabia, Mexico, Canada and perhaps Iran. Furthermore, this is only the beginning, because countries like Saudi Arabia and Iran can not only push themselves into the bulk chemical market — selling their products to firms that help them to establish their installations, or to brokers — but in the longer term they can also begin to produce the more sophisticated intermediate or final products. This will not happen soon, but when it does happen the countries thus involved will effectively have begun to break their dependence on petroleum.

WORLD STEAM COAL MARKET

Nineteen eighty-four was not only the first year in half a decade when, on a worldwide basis, the consumption of oil increased, but it was also a good year for the world coal market. Imports rose by 14 per cent during the year to 149 million tonnes (= 149 Mt), with seaborne trade rising by nearly a fifth to 104 Mt. Adding coking coal to this total means that total world trade in coal increased by 14.5 per cent to 305 Mt, with seaborne trade amounting to 235 Mt. Most forecasts suggest that trade in coking coal is on the verge of stagnation but that the sky is the limit for steam coal.

Whether this is true remains to be seen, because only a few years ago the predictions were that the steam coal market was essentially without a future until a great deal of capacity disappeared in some of the principal exporting countries. It may be that the high sales volumes and good prices that characterised 1984 were due to random events that will not return in the near future. Among these were favourable business conditions and a strong dollar that, because coal is largely priced in dollars, meant windfall profits (in their own currencies) for almost all exporters except the United States. It was a good year for South African and, especially, Australian exporters of coal. In 1984 Australian steam coal exports increased by approximately 56.5 per cent to 28 million tonnes and Australia became the top world coal exporter — where coal here means steam plus coking. 1985-6 has also been very satisfactory for Australia.

Only 50 years ago coal was the most important energy resource, as well as a basic raw material in the chemical, gas and metallurgical industries. Since then it has steadily lost ground to other fuels, and now supplies only about 21 per cent of the energy consumed in the

World Outside the Centrally Planned Areas (WOCA), and 30 per cent in the world as a whole.

The activity consuming the largest amount of coal is electricity generation, followed by steelmaking. Slightly more than half of the world coal trade is in metallurgical or coking coal. The recession has caused a sharp fall in coking coal prices and a downward trend in coking coal consumption in the industrial countries, but many observers believe that this trend will eventually be reversed. For example, the celebrated World Coal Report (Wocol) expected the import demand for coking coal to reach 260–300 million tons of coal equivalent per year (Mtce/y) by the turn of the century, but I think it can be said that this prognosis no longer has a basis in reality.

Where steam coal is concerned, the large scale intercontinental trade of this commodity is a relatively new phenomenon. Canada, South Africa, Australia, the USA and Poland are the most important exporters of coal, with most of Poland's trade taking place in Europe — where, in 1984, it sold 32.6 Mt of steam coal, which was a rise of more than a quarter on its 1983 performance. These countries may be joined in the not too distant future by Colombia, China and Indonesia; while in the distant future the Soviet Union, which has the largest reserves of coal in the world, should become a large exporter. It has also been suggested that, in the long run, Botswana might be able to become an important exporter. Between 1973 and 1980 the volume of seaborne steam coal increased by a factor of almost four, from 19 to 74 Mt, with the largest increase registered by South Africa. Expansion has continued in three of the past four years, with the exception of 1983, and as a result, for the first time in modern history, more steam coal than metallurgical coal is being traded by sea. Table 2.2 provides some information on the imports and exports of coal, and the reserves of coal.

Only a few countries have complete data on proven reserves of coking coal. About 20 per cent of the proven coal reserves of the United States are metallurgical coal. In Australia and West Germany, metallurgical coal accounts for approximately 60 per cent of total bituminous coal reserves. British and South African coal is mostly steam coal. In terms of carbonisation and calorific values there are three broad categories of coal: *anthracite*, accounting for about five per cent of production, *bituminous*, and *sub-bituminous* (70 per cent), and *lignite* (25 per cent). Lignite is sometimes called *brown coal*, and is characterised by a low calorific value. Sub-bituminous coals also have a low calorific value, and like lignite do not enter into world trade. Anthracite is a very hard coal, which

Table 2.2: Coal production, exports, imports and reserves

Country	Production[a] 1985	1984	Exports[b]	Exports[c]	Country	Imports 1985	1984		Reserves[e]		Consumption[f] 1985	1984
Australia	127	125	89	29	Canada	15	18	Western Europe	37,090	China	806	680
United States	800	809	82	22	West Germany	10	9	USSR and Eastern		United States	685	661
Soviet Union	726	712	—	13	United Kingdom	13	9	Europe	156,414	Soviet Union	551	510
China	847	771	8	4	Italy	22	20	Middle East	1,921	Poland	157	153
South Africa	172	146	43	30	France	21	24	Africa	42,824	South Africa	128	120
Canada	61	57	28	3	Denmark	13	10	Far East and		Japan	110	105
West Germany	89	95	9	3	Netherlands	12	10	Australasia	128,756	UK	102	79
United Kingdom	92	51	2	2	Spain	7	8	North America	167,855	Germany (FR)	86	86
France	15	17	2	—	Japan	92	86 (19)[d]	Latin America/				
Spain	—	40	—	—	South Korean	19	14	Caribbean	7,955			
Poland	192	192	36	33	Hong Kong	6	5	*World*	542,815	*World*	3,204	3,040
India	150	145	—	—								
South Korea	22	—	—	—								

Notes: [a] Production of steam + coking coal, in Mt/y
[b] Total exports, in Mt/y
[c] Steam coal only, in Mt/y
[d] Steam coal imports, in Mt/y
[e] For 1984, in Mtce
[f] Latest estimates

Source: Shell briefing reports (1984, 1985, 1986).

contains less volatile matter than bituminous and sub-bituminous coals, although its calorific value may be slightly lower than high-ranking bituminous coals. Almost all coal traded is bituminous.

Table 2.3 suggests the development of consumption patterns and the trade in coal. Many potential users of coal have refused to switch to that commodity until they have a clearer idea of its availability. This is particularly true of industrial consumers (excluding the iron and steel industry), which currently purchases 20 to 25 per cent of steam coal output. The same behaviour has been displayed by some suppliers of electricity — although there are several important exceptions. One is the Cincinnati Gas and Electricity Company which, in January 1984, elected to scrap a nuclear plant that was 97 per cent complete and replace it by a coal using facility. The reason given was that the cost of meeting new safety regulations would have required an expenditure roughly equivalent to the 1.7 billion dollars already invested. Furthermore, in The Netherlands, Italy and Belgium, a considerable amount of oil was replaced by gas in power stations. This has been termed an interim measure in these countries; but because of the increase in supplies of natural gas to Western Europe following the completion of the Yamal pipeline, the price of all gas in Europe has fallen at least as rapidly as the price of coal, while the long term availability of natural gas appears to have increased dramatically. The average spot price of steam coal in Europe fell from more than 50 dollars/tonne in 1982 to approximately 42 dollars/tonne at the end of 1984. At the same time average coal prices on medium and/or long term contracts may have fallen by between 15 and 20 dollars/tonne.

Table 2.3: Coal consumption (1985), and forecast consumption (1990, 1995 and 2000), coal trade (1985), and forecast trade (1990, 1995 and 2000)

Consumption (in tonnes)	1985	1990	1995	2000
Electric power (steam coal)	2,442	2,828	3,220	3,631
Industrial and other uses (steam coal)	1,109	1,203	1,288	1,390
Metallurgical coal	524	524	525	526
Synthetic coal	39	39	39	39
Trade (in tonnes)				
Metallurgical (total)	148	160	165	165
Metallurgical (seaborne)	130	142	147	149
Steam coal (total)	157	208	241	293
Steam coal (seaborne)	116	167	204	255

Source: Calculated by the author from International Energy Agency coal reports (1985), and Shell Briefing Service (1985, 1986).

This portion of the discussion can be concluded by reminding the reader that there are two basic methods of mining coal: underground and surface (open cast or open pit) mining. The production costs for surface mining are usually much lower than for underground mining, mostly due to the economies of scale that result from greater mechanisation. The lowest cost coking coal in the world from open cast mines is to be found in Queensland, Australia, while the lowest cost steam coal is from South Africa. The breakeven 'mine mouth' coal price for these installations appears to be roughly 15 dollars/tonne and 20 dollars/tonne, respectively, which means that South African steam coal can be delivered to Rotterdam at a price of about 45 dollars/tonne. The least expensive underground coal in WOCA probably originates in Utah, USA, where its mine mouth cost may not be more than 20 to 22 dollars/tonne; but rail and port charges add substantially to the free-on-board (FOB) price of this coal. Underground steam coal from New South Wales is also inexpensive, and its cost at Australian coal terminals is approximately 40 dollars/tonne. This price situation is probably responsible for Australia's palpable progress in its bid to overtake South Africa and Poland, which are the two countries still ahead of Australia in steam coal exports.

United States steam coal and Western Europe

The desirability of increasing exports of United States steam coal to Europe was advanced well before the first oil price shock. In a talk to the Colloque Européen de l'Energie held at the University of Grenoble in May 1965, Professor Morris Adelman reviewed some of the issues involved in making American coal a more prominent part of the Western European energy picture. Professor Adelman's principal arguments turned on European energy security.

A dynamic upgrading of United States shipping facilities commenced in the late 1970s, and indeed an export boom did begin in 1978. Unfortunately this boom peaked in 1981, and since that time United States coal exports have been in retreat. Signs of trouble had been on the horizon for some time, however, with a fall in the rate of growth of electricity consumption throughout the world. For this and other reasons there is good reason to believe that the gusto has gone out of the European coal market for the rest of this decade as far as United States suppliers are concerned.

Before presenting an account of what I believe to be the basic long run dilemma for the United States coal industry, a caveat will be offered, since undoubtedly a great deal more United States coal is going to be sold in Western Europe. The largest colliers (coal carriers) range in size from 100,000 to 200,000 deadweight tons (dwt) or more; and when fully loaded the larger vessels will draw over 50 feet of water. This is deeper than the mean high tide of nearly all United States ports; and at present no major port on the East Coast is deeper than 45 feet. Although the approval process for dredging is almost grotesque in its complexity, the dredging required to create at least one deep-water port on the United States eastern seaboard will take place sooner or later. At present, about 60 per cent of internationally traded coal is carried in ships between 40,000 and 80,000 dwt, and only seven per cent in vessels of 100,000 dwt or more, even though economies of scale lead to a 50 per cent reduction in costs when the size of colliers is increased from 60,000 to 150,000 dwt. Although it is difficult to be precise in these matters, my own calculations indicate that when larger colliers come into more general use, and if ocean transport costs rise from their present low levels, the competitiveness of United States coal in Europe will increase relative to that of coal from South Africa, and even more to that of coal from Australia.

The railroads in the United States — and, if not the railroads, Congress — may eventually come to the view that sound arguments exist for substantially reducing coal haulage rates: if these rates fall then low cost coal from the west of the US (and perhaps even the midwest) would be favoured. Furthermore, this coal could be obtained from larger mines or deposits than generally exploited in the east, and thus western American firms could sign longer term contracts with consumers than, for example, eastern firms. This would mean that — unlike producers selling on spot or short term contracts — the shipments of these large mining firms would be less influenced over time by the 'prevailing' price, and to a considerable extent even by falling demand. Everything considered, it might be possible for the railroads to more than make up in volume what they lose in price. There are also the diversification arguments in favour of United States coal that are well known to almost everybody except academic economists, who insist on concentrating on prices instead of also taking into consideration quantities, uncertainty and contracts that can run 20 years or longer. Finally, bunker fuel prices could rise in the long run, which would favour the United States more than Australia, and to a certain extent South Africa.

If even some of the above is true, then what is the problem? Why should paradise elude the sellers of United States steam coal, and their prospective, eager, and often politically motivated customers? The reason is that, despite its enormous reserves and the present glut of coal, there are uncertainties among potential buyers about just how much 'reasonably priced' coal the United States is capable of supplying in the medium or long term, without a dramatic modification in the present geographical pattern of production. The expansion in United States exports and production mentioned above was accompanied by a substantial increase in United States coal prices — an increase that could hardly have been avoided given the part played in this drama by hundreds of small and medium-sized mines in the east and midwest, whose short run costs often rose more than proportionally to output, and where scale economies are virtually unobtainable at the present level of mechanisation. In addition it appears that there could be further substantial rises in the cost of obtaining low sulphur Appalachian steam coal not too much further along in the depletion cycle.

Similarly, the cost of medium-sulphur midwestern and Appalachian coal could also increase, though not so strikingly as for low sulphur coals; but this is irrelevant since Western European importers try to avoid buying medium sulphur coal for environmental reasons. In these circumstances, and considering the matter of depletion mentioned earlier, it might be some years before the United States can be regarded as the Saudi Arabia of coal: that is, the repository of enough high quality, cheap coal to mitigate the discomforts that could be visited on Europe in the event of a new and sustained energy price rise.

Pacific Rim

When the Pacific Rim is mentioned, attention swings immediately to Japan, which imports about a quarter of the coal entering into international trade. As it happens, though, about 80 per cent of these imports are of metallurgical coal, which constitutes the main part of Japanese coal consumption. Domestic Japanese coal output has fallen from 55 Mt in 1961 to about 18 Mt in 1982, but recoverable reserves of 1.1 Gt should permit the present level of domestic production to be maintained throughout the rest of this century. The Ministry of Industry and Trade recently estimated that Japan will be using about 110 Mt of coal in 1990, with 45.3 to 48.3 of this being steam coal, instead of the 60 Mt predicted just two years earlier.

Similarly, the Japanese Institute of Energy Economics estimates a 1990 demand for steam coal of 39 to 44 Mt. The Japanese have had extraordinary success in reducing coal-based emissions, and so pollution is less of a constraint on the use of coal in that country than in North America or Western Europe; however the security problem associated with managing nuclear energy may be less complicated in Japan than elsewhere, and so there is considerable pressure from decision makers in government and industry to use as much nuclear energy as possible. Here it should be noted that the accident at Chernobyl does not seem to have had the slightest effect on Japanese nuclear plans.

Japan is going ahead with the building of large power stations, and decisions seem to have been made to ensure the commitment of Japanese talent and technology to the development of such things as coal liquefaction, fluidised bed combustion, technologies for burning coal water mixes and coal gas mixes, and so on. At the same time Japan is to be heavily involved in the rapidly expanding Chinese coal sector.

South Korea, which also produces coal, will probably be a major importer of steam coal by the year 2000, with much of this coal coming from mines in Canada and Australia in which there are large South Korean interests. Some observers believe that South Korea is on the verge of a major switch to coal, with coal demand rising to between eleven and 14 Mt by 1992, and 35 Mt by the turn of the century. Smaller amounts of coal will probably be imported by Taiwan, Singapore and Hong Kong — although there is very definitely a possibility that Taiwan will be a major importer. The cost-minimising supplier of coal to the Pacific region is clearly Australia, whose main coal resources are located in New South Wales and Queensland. Surface mining accounts for about 55 per cent of Australia's total output, and the geology of open cast mines in Australia is roughly comparable to that of low cost open pit mines in the United States west. A World Bank report by Choe (1985) implies the cost of coal produced in Australia increased by almost 20 per cent between 1977 and 1982, mostly because of increases in wages and salaries; but even so in 1983 the FOB cost of steam coal ranged between 35 and 55 Australian dollars per tonne, which was not particularly high. (In 1984 spot steam coal rose by a total of 12 dollars per tonne.) Moreover, within this cost range, substantial new capacities can become available, which means that the average long term breakeven price for Australian coal may be considerably under that of coal from the United States unless the value of the US dollar continues to drift downwards.

At present Japan buys about half its steam coal from Australia, and it is generally believed that this arrangement will continue over the foreseeable future. However it is an open secret that the Ministry of International Trade and Industry (MITI) favours stronger economic links with the United States, and if possible wants Japan to buy more coal from the United States — even though a price might have to be paid for this coal that is at least ten, and occasionally as much as 15 per cent higher than the Australian price. For instance, in 1984 the average CIF price for coking coal imported into Japan was 63 dollars/tonne, while it was 71 dollars/tonne for United States coking coal. The average price paid for Australian coal during the same period was 59 dollars/tonne (all these quotations are in United States dollars).

Exactly how the United States coal industry and the United States government, will eventually respond to this overture is difficult to say. The cost of hauling coal by rail will undoubtedly be reduced but as things now stand this cannot be done quickly. Even with unchanged rail costs, if the United States West Coast ports of Portland and Long Beach had the same capacity as, for example, Abbot Point in Queensland, then the delivered price gap in Japan between United States and Australian coal could almost be halved. The question is, however, whether this would enable United States companies to sell more coal in Asia than they were going to sell anyway, even if their prices are substantially higher than those of competitors. True, the Japanese want to diversify as widely as possible, but it should be remembered that the Japanese also have access to South African and Canadian coal, as well as Australian coal. As pointed out earlier, South African steam coal is the least expensive in the world FOB Richards Bay; and when the domestic transport system in Canada is better developed — and this is taking place now — the cost of some Canadian coal could come down to the point where it can be delivered to Yokohama for the same price as Australian coal. The average rail transportation distance from mine to port in Western Canada is about 700 miles, and coal loading facilities near Vancouver can handle ships of 150,000 dwt.

For geographical reasons, the natural supplier of coal to Japan may be China. The rate of increase of coal production in that country in the past 25 years has been the highest in the world and, as far as I can tell, the Chinese are not many years away from developing a substantial export capacity. Similarly, once the new trans-Siberian railway — which in large part runs parallel to the original line — is completed, the Soviet Union should be able to export considerable amounts of coal to Japan. This will not be soon, however, and most observers do not expect this trade really to get underway before late this century or early in the next.

I will conclude this discussion with a few general points. Given the mountains of unsold coal that can be found in virtually every part of the world, investment has been inhibited in most of the traditional coal-exporting countries; but in several countries with important coal resources — Poland and Colombia — serious attempts are being made to increase coal exporting capacity as rapidly as possible. Regardless of environmental costs, attempts are being made in Poland to produce and use more brown coal in order to free larger amounts of high quality bituminous coal for export. The prediction here is that these attempts will be successful if the political situation in that country does not deteriorate, and given present trends, Poland will succeed in capturing new markets in Western Europe, as well as expanding old markets, at the expense of all the major coal exporters. By the same token, with Colombia capable of overachieving its export goals for 1990, there will probably be no additional place in the world coal picture for a resurgence in United States exports — especially if Colombia exports in excess of 35–45 Mt by the end of the century. Here it should be noted that coal from the main Colombian mine, El Cerrejon, although labelled steam coal, can be regarded as a 'semi' coking coal, and if blended with good quality coking coal can generate considerable cost savings. The first deliveries from El Cerrejon have taken place ahead of schedule, and unless political trouble stops the additional tonnage that should be available from South Africa, it will be some time before there will be the sustainable upward pressure on the price of coal that all coal producers have been dreaming of over the last decade (for more on these topics see R. Steenblik, 1983, 1985).

NATURAL GAS

As the statistics in the previous chapter indicated, and figures presented later in this book support, natural gas consumption and production now seems to be on an upward trend. Some observers have even gone so far as to signal the start of a new energy era. The price of natural gas is falling, which keeps most producers from agreeing with these observers, but a declining price might well have a beneficial long term effect because it could allow natural gas to penetrate new markets, maintain existing markets and encourage the building of pipelines and/or LNG facilities. Among other things this could mean that consumers were willing to commit themselves to buying very large amounts of gas over a very long time horizon at gas prices that were in line with energy prices in general.

Table 2.4 shows that trade in natural gas increased in volume between 1983 and 1984, and this tendency has been maintained up to the present (mid-1986). Trade is expanding faster than production, although the traditional pattern is being maintained of using most of the gas produced in the vicinity of the place of production. Changes, however, can be expected. In 1980 natural gas in Western Europe accounted for an 18 per cent share of total delivered energy; but by the turn of the century this figure will be at least 22 per cent of a much larger energy demand.

Table 2.4: World trade in natural gas by importing countries (1984–5)

		Gcm	
		1985	1984
Japan		37.5	33.2
Western Europe		127.8	110.8
Of which:	West Germany	44.1	34.1
	France	25.6	21.3
	Italy	19.7	18.7
	United Kingdom	13.6	13.2
	Belgium	9.1	10.0
America		27.4	25.1
Of which:	United States	25.1	23.0
Eastern Europe		39.1	34.2
Of which:	USSR to Eastern Europe	36.4	31.8
World total		232.6	219.0
Of which:	Liquefied natural gas	51.2	48.0
	Pipeline	181.4	171.0

Source: Calculated from basic data in *The Petroleum Economist* (1983, 1984, 1985)

As part of the explanation for why there will be an expansion in the production and trade in gas, it should be noted that in Europe the industrial and power sectors currently consume about 74 Gcm/y (= 63 Mtoe/y = 1.26 Mbbl/d) and 31.6 Gcm/y (= 27 Mtoe/y = 0.54 Mbbl/d) respectively. Within the industrial sector the chemical industry is the most important consumer, using about 40 per cent of all industrial gas. If European industrial activity increases at the expected 2.5 to 3 per cent/year rate, it means that by the year 2000 industrial consumption of gas might reach almost 87.8 Gcm/y (= 75

Mtoe/y = 1.5 Mbbl/d). The way this figure is obtained is to observe that European industry used a total of 250 Mtoe in 1984, of which approximately 25 per cent was natural gas. Assuming that each percentage point increase in economic growth results in a percentage point increase in energy use, we get the above figure. Where power generation is concerned, natural gas may regain the competitive position lost during the early 1980s to heavy fuel oil, but at the same time it could become more expensive with respect to coal. However, given the present attitudes in Western Europe toward the environmental destruction that may be caused by coal, as well as the fall in gas prices that could take place if oil prices continue to slump, gas should do better in the coming years relative to coal than once thought. The consumption of gas may also increase relative to the consumption of nuclear power in the power generation field.

We next go to a brief overview of the world supply-demand situation for natural gas, by region, beginning with Western Europe. In 1976 The Netherlands supplied 47 per cent of the European consumption of natural gas, but by 1983 was supplying only 33 per cent. The United Kingdom and Norway together also supplied 33 per cent, while European imports were 16 per cent of its consumption. With local production remaining essentially constant, but total consumption increasing, imports are going to become more important, and especially imports from the Soviet Union and Algeria. At the present time Soviet gas is generally judged less secure than Norwegian, Dutch or even Algerian supplies, although contracts signed by the Soviet Union seem more flexible than most, and as far as I know all customers of that country (with the possible exception of Iran) have expressed satisfaction with the businesslike attitude of the Soviets.

The key to Soviet reliability is its need for new sources of foreign currency as its reserves of oil inexorably diminish, and it appears that the Soviet Union may be prepared to sell large amounts of gas at prices that are well under those that competitors and potential competitors are willing to offer. The Soviet Union already has two pipelines that reach Western Europe and which could, under ideal conditions, deliver 80 Gcm/y (= 65 Mtoe/y), which is almost 25 per cent of Western Europe's projected natural gas demand in the year 2000. In July 1985 the Soviets agreed to reduce the price it charges France for gas delivered under two major contracts, and it is estimated that this revision will save France one billion francs on its 1985 gas bill alone. Several weeks earlier the Soviets agreed to reduce the price on a 1982 gas supply contract to 3.5 dollars/MBtu, and at the same time the president of Soyuz Gazexport said that gas will replace oil as the Soviet Union's principal earner of foreign exchange in a few years. My personal expectations are that the Soviet Union will eventually become the largest supplier of gas to Western Europe, and will display at least as high a degree of reliability as other suppliers. Here it should be noticed that the Dutch policy on export and production levels has vacillated considerably over the past half decade or so, but recently it was announced that since larger reserves of natural gas are available in The Netherlands than previously believed, export contracts will be extended.

The situation in the North Sea will be taken up in considerable detail later, but a few observations are in order now. The United Kingdom is the second largest West European gas producer, but is already taking imports from Norway's Frigg Gas Field. The British Gas Corporation believes that a great deal more gas will have to be imported in the 1990s, and wanted to contract for about 30 billion dollars worth of gas from Norway's Sleipner Field. The British government, however, claimed that reserves in the British North Sea were underestimated, and in addition seemed to think that if the price of gas were to rise, new supplies would be forthcoming from indigenous resources. Accordingly, they proceeded to cancel the proposed British Gas-Norwegian deal. The price provisionally agreed upon for Sleipner gas was 4.10 dollar/MBtu, which is well above the 3.10 dollars/MBtu at which domestic UK was selling, and presumably the government believed that a gas price of 4.10 dollars/MBtu would increase the availability of domestic gas. At least one company operating in the North Sea believes this is false, and the opinion here is that sooner or later the British Gas Corporation will be allowed to contract for additional gas, though perhaps not so much as was going to be purchased under the now cancelled Sleipner agreement. Increasing mention is also being made these days of a possible link between the United Kingdom and the Continent, with the United Kingdom receiving gas from The Netherlands, Algeria (via France), and the Soviet Union. The latter possibility is looked upon with great distaste by, among others, the present United States government, which wants to see Norway given priority over other potential suppliers. It has been suggested though that a link with The Netherlands or France would enable the United Kingdom to arrange for spot supplies from various exporters (to include the Russians), and thus if the UK North Sea does not prove capable of supplying more gas at higher prices, the United Kingdom will still have a facsimile of the security that was to be obtained by signing long term contracts for large quantities of gas with Norwegian suppliers.

Norway was not happy about losing the contract for Sleipner gas but appears to have altered its previous sales philosophy. Now, instead of setting a definite annual production target for gas, the Norwegians seem prepared to wait until the right offer comes along, and when (or if) it does, offer entire fields or large parts of a field for sale. The right deal now appears to be selling 68 billion dollars of its gas over 30 years to a European consortium led by Germany's Ruhrgas, and the agreement covers sales to West Germany, France, Belgium and The Netherlands. Eventually the Norwegians hope to sell to Italy, Greece, Spain and also Sweden. In the process they hope to double their share of continental gas exports to about 25 per cent by the year 2000. This can be compared to the situation for the Soviets, who now have a 30 per cent market share. The reserve/production ratio for Norwegian gas is about 100, and the Troll field alone is probably able to supply present Norwegian markets for more than 50 years. The present deal lumps gas from the Sleipner field with that from the Troll field, and will involve volumes of 607 Gcf/year. Deliveries are scheduled to begin in 1993, when Sleipner will come on line.

At present Algeria is expected to make a major contribution to Western Europe's energy future. Large quantities of Algerian gas have already been contracted for, and at the beginning of 1985 Algeria had the capacity to export about 47 Gcm/y (= 40 Mtoe/y) by pipeline or in the form of LNG, although only a quarter of this is now being sold. By the year 2000 Algerian exports are predicted to be in excess of 35 Gcm/y. For a long time Algeria insisted that its gas must be priced at a level equivalent to crude oil or even synthetic gas or oil calorific value, but like other gas suppliers was forced to alter this position when huge new supplies of Soviet gas appeared on the market. For two years Algeria refused to fill the trans-Mediterreanean pipeline, but in 1984 Algeria and Italy signed an arrangement to enlarge the capacity of the pipeline, and deliveries are scheduled to increase from 7 Gcm in 1984 to 12 Gcm by 1986 under a 25-year supply contract. Norway was one of the other suppliers of gas that was forced to alter its expectations of future gas prices, and an important question for both these countries — as well as for all other actual and potential consumers and suppliers of gas — concerns what could happen to the price of gas if the huge supplies of Iran could be brought to market.

Algeria is a member of OPEC and the most important Arabic

producer of natural gas. There is also considerable gas in the other Arab states belonging to OPEC, but these countries are in a more favourable economic position than Algeria because they have a great deal of oil as well as smaller populations, and feel no compulsion to lift their gas. Much of the gas of Algeria is non-associated, while much of the gas in the Middle East is associated, and the countries possessing it have elaborate plans to harness it. Just now Saudi Arabia is routing almost all its associated gas into a huge gas gathering system, and using this gas in petrochemical installations, cement and glass plants, and in power stations. This gas can also be used as the energy source for refineries, which means that even if Saudi refineries were compelled to pay the 'world price' for the crude oil they will process, they could still produce refined products (FOB factory) cheaper than most refineries elsewhere in the world. It also seems likely that before the end of this century more major LNG projects will originate in the Arab countries of the Middle East and Iran. Qatar, for example, is already a very important producer.

Asia is a part of the world where natural gas markets are expanding, with Japan as the focal point of consumption. The newest projects consist of LNG deliveries from Indonesia and Malaysia to Japan, but Japan will also start receiving Australian gas in the not too distant future. The Japanese have also shown considerable interest in buying gas from Canada and North America. Their Australian links consists of the North West Shelf project which was initiated by Woodside Petroleum, and which is partly owned by Shell and BHP. The latter two firms have a quarter share in the export phase of the project, with a sixth of the shares held by BP, Chevron, and a partnership of Mitsubishi and Matsui. The proposed Canadian project is not as large as the Australian, and may not be realised in the near future. But even so there are still a large number of Japanese firms which hope eventually to import a great deal of Canadian LNG. Similarly, 16 Japanese firms have been discussing the possibility of obtaining 730 Gcm of LNG from Prudhoe Bay in Alaska, and with New Zealand gas production increasing again, and new fields apparently about to come on stream in the Gulf of Thailand, it is clear that the price of gas to Japan is unlikely to increase. Even with the high rate of growth of gas consumption in the world, reserves are increasing more rapidly; and the increase in the world reserve-consumption ratio can only put further downward pressures on the price of gas. Since the Japanese understand this situation perfectly, they are in no great hurry to sign long term contracts with many potential suppliers, and probably regret some of the contracts they have already signed, and in addition would like to retrieve some of the money they have invested in energy proejcts around the Pacific Rim and elsewhere. For instance, it seems almost certain that if Japanese energy importers actually possessed the perfect foresight we like to claim for economic agents in our textbooks, they would not have invested any money in Australia's Northwest Shelf.

According to the latest plan, Japan's LNG requirements are scheduled to grow from 27.3 Gcm in 1984 to 51 Gcm in 1990, and 56 Gcm by the year 2000. South Korea and Taiwan have also expressed considerable interest in becoming major users of natural gas, almost certainly in LNG form. An important supplier to Japan that was not mentioned above is Brunei, which is able to support a major export programme with less than 0.2 per cent of the world's natural gas reserves. At present there are 14 countries in the Pacific region producing gas, and over the past two years all of these have registered an increase in production with the exception of Taiwan. Mainland China is increasing its production at a fairly rapid rate, although that country has not indicated an interest in exporting gas. India expects to become a large producer in the not too distant future, and Japanese technicians have discovered a field off the Burmese coast. A gas condensate field has even been discovered in the interior of Papua New Guinea. As far as I know, the expectations that existed a few years ago concerning the discovery of a super-giant offshore field in the vicinity of Malaysia have been dampened, but not abandoned. Interest also surrounds the possibility of a joint Japan-USSR exploitation of gas reserves in the vicinity of Sakhalin Island. A major question here concerns the exact extent of the reserves, but if there is as much as some people think, this gas will provide a valuable addition to the Japanese energy picture.

This review is concluded with the Americas, beginning with the United States. Although half the world's offshore gas is produced in that country, total gas production began falling in 1983, and despite a recovery in 1984 to 487.5 Gcm (= 17,216 Gcf), the United States does not have the inherent production capacity of a decade ago. It has been claimed that the principal reasons for this are low well-head gas prices (which average about 2.85 dollars/MBtu or 2.95 dollars/Mcf) which have caused drilling to fall off; but even so it is clear that at present levels of demand there is spare capacity on the production side of the United States gas market that the United States Department of Energy has estimated at 25 per cent, and which could persist until the last part of this decade. By that time many observers expect the gas bubble (or excess supply of gas) to be replaced by a gas shortage, with rapidly rising imports. Although gas company executives have argued that the decontrol of gas prices will lead to increased exploration and drilling, the simple truth of the matter is that there is much less easily locatable conventional natural gas waiting to be found in the United States than gas consumers and producers consider desirable, and this situation cannot be altered by just raising the gas price. At the same time, however, it might be true that a higher gas price will promote the technological developments needed to speed the discovery and exploitation of 'unconventional' natural gas. This is gas situated at much greater depths than the gas that is being found today, as well as gas in unusual and/or exotic environments both on and offshore.

The export of Canadian and Mexican gas seems to have stagnated, but this is largely because of wavering demand in the United States. The domestic price of Canadian gas is fairly low — about 2.10 dollars/MBtu — while the government established export price for the last part of 1984 was 4.40 dollars/MBtu for base volumes, and 3.40 dollars/MBtu for excess volumes. These latter prices were, however, reconsidered by the government because of their desire to expand the foreign market for Canadian gas, and so called 'arms length' negotiations became possible for Canadian gas producers. In practice this could mean border prices as low as three dollars/MBtu. On the other hand Mexico has indicated an unwillingness to sell its gas at lower prices, and hopes to compensate for the revenue decline that this will mean by selling more oil — if possible. Present plans call for more gas to be allocated as a petrochemical feedstock and as a fuel for industry and utilities.

Finally, in Latin America and the Caribbean, gas production is up in Argentina, Venezuela and Trinidad. Brazil would also like to use more gas as a petrochemical input, and to replace fuel oil in industry and power generation, but whether this will be possible depends upon whether Brazilian expectations regarding the discovery of gas within the borders of their country will be fulfilled, and this is far from certain. The decision makers in Colombia have been thinking along the same lines, although the recent large oil discoveries in that country will probably call for a change in plans. Argentina is basically self sufficient in gas, and is extending its onshore gas distribution network in order to facilitate greater consumption of gas. There has also been talk of pipelines to carry gas from Argentina to Brazil (Sao Paulo) and Chile. Given the opportunity, Argentina would like to introduce compressed natural gas as a motor fuel, employing the same technology that has given such satisfactory results in New Zealand.

APPENDIX: RESERVE/PRODUCTION RATIO

The reserve/production (R/P) ratio is one of the most important concepts in petroleum economics, but one of the least understood. It can be approached as follows.

In order to maximise the amount of oil that can be taken from a deposit, no more than a fraction of recoverable reserves should be removed in a given year. This fraction is generally put at about one-tenth. Assume that we have a field containing 150 units of oil, and we desire to lift 10 units per year. Since 10 is less than one-tenth of the reserves of this field, we can remove 10 units/year for 5 years (measured from the end of the year) without violating the above criterion. During this time the R/P ratio falls rom 15 (at the end of the first year) to 10. Some simple algebra shows this situation. Taking Q as production, and R reserves, we must have $Q = -(dR/dt)$. This expression can be integrated to give:

$$R(T) = R(O) - \int_0^T Q(O)e^{nt}dt = R(O) - Q(O)T.$$

The right-hand side of this expression was obtained because of the assumption (in this example) of constant reserves, and n (the rate of growth of consumption) equal to zero. Note also that in this example $Q(T) = Q(O)$. Now, multiplying the first expression on the right-hand side, $R(O)$, by $Q(O)/Q(O)$; and then dividing $R(T)$ and the right-hand side by $Q(T)$, we obtain in a straightforward fashion:

$$\frac{R(T)}{Q(T)} = \frac{R(O)}{Q(T)}\frac{Q(O)}{Q(O)} - \frac{Q(O)}{Q(T)}T = \frac{R(O)}{Q(O)} - T.$$

In the numerical example above $R(O)/Q(O) = 15$, and when $T = 5$ we have $R/Q = 10$, which is the critical R/P ratio. Observe that $R/Q = 15$ at the beginning of the first period, but 14 at the end (or, equivalently, the beginning of the second period).

After the fifth year, the critical R/P ratio takes over and determines production. In the sixth year reserves are 90 units, and with a critical R/P ratio of 10, production cannot be larger than 9. In the following year, with reserves 81, production can only be 8.1 (= 81/10), and so on. Something that should be pointed out here is that in a country where fields have different sizes, and where some fields are under development and are not producing oil, the critical R/P ratio may be more than that of an individual field. Readers interested in this matter will find a simple example constructed in *The Political Economy of Coal*.

A mathematical analysis in which $n \neq 0$ and reserves grow over time can be found in the appendix to Chapter 10. Future research, however, will probably not be particularly concerned with complicating the above model, but with disaggregating the R/P ratio across groups of countries. For example, the important thing about this ratio at the present time is not that it is about 35 on a worldwide basis, but that outside OPEC and the centrally planned areas it is only about 18.

3

Economic Theory and Natural Gas

Most of the technical content of this book can be found in its penultimate chapter, but there are a few simple concepts that must be introduced in the course of a fairly general discussion about the markets for natural gas.

In this chapter I consider pipelines, liquefied natural gas and offshore gas. As the reader probably understands by now, the first step in the gas supply process involves finding the gas: removing it is a fairly simple technical operation. Whether the gas is located together with or independent of oil, it will be under pressure, and the presence of an outlet (in, for example, the form of a pipe driven down into the deposit) will cause the gas to leave the deposit under pressure.

As the gas flows out, this pressure decreases. If the gas is to be used only a short distance from the deposit, a decrease in gas pressure may not be of great importance: almost all the gas will have left the deposit before the pressure falls far enough to prevent the gas from being captured and routed to its purchasers by natural means — i.e. without the use of compression equipment. But if the gas must flow a large distance, then equipment in the form of compressor stations must be inserted in the gas circuit to ensure that the pressure is high enough to push the gas to its destination, and to give it the required pressure at its destination (which in many cases is a local distribution system). In some cases compressors are not needed at first but are installed as pressure drops or the demand for gas increases (Figure 3.1).

Figure 3.1: Simple flow diagram of gas transmission process

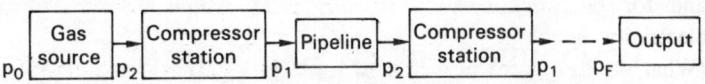

The pressure at the source is p_0. This eventually falls to p_2, and is raised again to p_1 by the compressor. (Note that there is a pipeline between the source and the first compressor station. As is often the case this can be a long pipeline because of the pressure at which the gas leaves the source.) Because of such things as friction losses in the pipeline the pressure falls again, and is once more raised by another compressor station. And so on. Outlet pressure is p_F.

As an example, I can cite the pipeline from Australia's North West Shelf to Perth, the capital of Western Australia. This pipeline is 26 inches in diameter and runs 940 miles, and there are five compressor stations through which the pipeline runs, as well as at least 50 mainline valves. Both the compressor stations and the mainline valves are controlled from Perth by a computer that can monitor on average 1,000 readings per second.

A similar project is the proposed gas line from northern Norway to the southern tip of Sweden, and from there over to Denmark and/or Germany. The line would be over 1,700 kilometers in Sweden alone, and the approximate cost would be 30 billion Swedish kronor, or more than four billion dollars at the present rate of exchange. This line would be 1.5 meters in diameter (which is almost 50 inches), and the pipeline wall would be 1.6 centimeters thick. Each length of gas pipe is 12 meters, and two lengths would be welded together into a 17-ton package before being put into the trench. In northern Sweden the trench would be 4 meters deep, and in southern Sweden about a meter. The idea is that the pipe should be maintained at a temperaure of 6 to 7 degrees Celsius. For reasons that are well known to students of fluid mechanics, at this comparatively low temperature there is room for a maximum amount of gas in the pipe; and if the pipe temperature is held more or less constant, the gas will not expand or contract.

A portion of the line would go through heavily forested districts, and in these areas an avenue about 20 meters wide would be cleared to facilitate the laying of the pipeline. Rust, which is a serious problem in pipelines, would be dealt with by a combination of plastic insulation and a weak electrical current along the outer surface of the pipe. Compressors would be spaced every 150 to 250 kilometers, and for cost minimisation reasons these would be driven by electricity, in contrast to the gas that is mostly used for this purpose. (What we have here in a limited sense is a partial transformation from electricity — which would probably be generated in nuclear installations — to gas.) Each compressor station would be about as large as a small industrial building, and would be staffed with personnel provided with transport and equipment to make repairs on the line. If this project becomes a reality, some of the gas would be used in Sweden. The project would also be a very large generator of employment in Sweden. It has been estimated that 4,000 people would work on the construction of the line, and another two or three thousand would be employed in enterprises providing materials and machinery for the pipeline and compressor stations.

The problem of determining the optimum number of compressor stations cannot be taken up in this elementary discussion, although in general it may be said that if the objective is to present a certain amount of gas at a certain point in space, then the factors of production are line pressure and pipeline diameter. The larger the diameter of a pipe, the smaller the amount of pressure required to move gas from one point to another — up to a point. Or, to turn this around, with a given line pressure there is a certain pipeline diameter that moves a given amount of gas from one point to another at minimum cost. The key expression here is 'up to a point'. Obviously, if gas has to be transported 5,000 kilometers in a pipeline, it will be impossible to compensate for a lack of compressors (and thus pressure in the line) by increasing the diameter of the pipeline.

It should also be appreciated that there are important economies of scale to be gained by increasing the diameter of the pipeline but these economies will not be realised unless the pipeline delivers a sizable amount of gas. Giving a precise explanation of why this is so would involve a technical discussion that I do not want to employ in this chapter, but the reader should be able to understand, intuitively, that if there is only a small amount of gas to be delivered, there is no point in using a large-diameter and relatively costly pipeline. With a given line pressure, increasing the amount of gas to be carried will often call for increasing the diameter of the pipe; however the increase in the cost of materials for a larger diameter line is usually modest, and with the other costs fixed, if a significantly large amount of gas is transported unit costs tend to fall. In a limited sense this is a classic problem. The carrying capacity of a gas pipeline is proportional to its interior cross-sectional area, which is πr^2, where r is the radius of the pipe. The amount of pipeline material (e.g. galvanised steel) used is proportional to the circumference of the pipe, which is given by $2\pi r$. If the input of steel is halved, the carrying capacity of the pipe falls to less than half the original amount; while if the input of steel is doubled, the carrying capacity is more than doubled. Obviously, however, there are limits to the scale economies that can be achieved by increasing the diameter of the pipe, and these limits are determined by — among other things — pressure differentials, the length of the pipeline etc.

It is more expensive to transport gas by pipeline than to move oil in a pipeline. In fact, one of the reasons why oil is generally called the highest quality source of energy is the relative ease with which it can be transported. Moreover, the advantage held by oil seems to increase as the distance over which the energy resource to be transported increases. The topic of LNG will be taken up later in this chapter, but the reader should remember that when gas is to be shipped by sea, it must first be liquefied, and later deliquefied, while crude oil and oil products can be pumped directly into a tanker. The capital investment for gas liquefaction and deliquefaction facilities can run into the billions of dollars, while no analogous outlay is required for shipping oil, or for that matter coal.

It is likely that there will be a considerable number of natural gas pipelines constructed around the world between now and the end of the century, since the annual increase in world energy consumption during this period is estimated to be between 1.5 and 2.5 per cent. It is impossible to pinpoint the world demand for gas by the year 2000, but it is conceivable that this demand will reach 2,200 Gcm (as compared to about 1,600 Gcm in 1983). This implies that more than 50 new major pipeline projects will be commenced in this period in the Soviet Union and Eastern Europe, the Middle East and Western Europe. Not much is expected from the United States in the way of large new projects, because gas consumption may fall in that country; but this forecast does not preclude major extensions to existing pipeline systems that could be extremely expensive. For obvious reasons pipelines do not have much future in the flow of gas from Australia, Brunei, Malaysia, etc. toward Japan, although there should be some major pipeline construction within Australia.

Western Europe's extensive pipeline network is already complete in many respects, which facilitates the switching of gas from one part of this system to another. But Western Europe will need much more gas in the future, and new lines will be needed to carry gas from the North Sea and the Soviet Union to the most heavily populated districts in Western Europe. In much of my previous work (e.g. *The Political Economy of Coal*) I have expanded on the desirability of bringing Middle East gas to Europe, and here I note two possibilities. One of these involves Iran resuming its export of gas to southern Russia, while the Soviets ship more gas to Europe; or even pipelines that carry gas direct from Iran through the Soviet Union to Western Europe. The other possibility is a pipeline from the Gulf area to Turkey that crosses the Bosphorus, and proceeds to Yugoslavia and Western Germany. Personally, I have no doubt that one or both of these pipelines will eventually be constructed, though perhaps not in the near future.

As implied above, it is generally believed that pipeline construction must decline in North America. The major portion of pipeline construction in that part of the world, in the future, will probably be in conjunction with efforts to bring Canadian natural gas to the United States. An example that is applicable here is the proposed 1,333 mile 36-mile pipeline designed to carry gas from the northern part of Canada to the southern Canadian and United States markets. The cost of the project has been estimated at 3.5 billion Canadian dollars, and the initial gas volume to be transported is 22.7 Mcm/day. This capacity can be expanded at a relatively modest cost if extra compressor stations are constructed.

The North American pipeline system, and particularly that of the United States, is now distinguished by its maturity, and the construction of additional line is seldom attractive from the economic point of view. There is, however, a considerable amount of looping going on, and so I will take the present opportunity to explain the meaning of this important term. Looping means adding pipe in parallel onto an existing gas (or oil) transmission pipeline. This is done when the pipeline pressure might fall below an acceptable level at some point in the line on the occasion of additional gas being fed into the line. Where typical pipeline costs for the United States are concerned, the following figures indicate the situation in 1983: all construction costs for an 8-inch pipeline were between 100,000 and 150,000 dollars/mile. This includes all the laying costs and the cost of the pipe. For a 20-inch diameter line costs came to 250,000–300,000 dollars per mile; and for a 36-inch pipeline 650,000–750,000 dollars/mile. Operating and maintenance costs are usually low. In the United States these costs were only about 7,000 dollars/mile/year, including the cost of gas used for compressor station fuel.

Nearly 60 per cent of these costs were related to gas metering and regulating. An important new pipeline carrying gas diagonally across India has the following projected costs: 600 million dollars for pipe 18 to 36 inches in diameter; 250 million dollars for laying the pipe; 250 million dollars for compressor stations; and 65 million dollars for telecommunications.

The country in which the largest pipeline projects are underway is probably the Soviet Union. The 2,030 mile (= 3,250 kilometer) 56-inch pipeline from the super-giant Yamburg field north of the Arctic Circle in Western Siberia to Yelets, southeast of Moscow, is to be extended to the Romanian border, and possibly on to Greece and Turkey. In addition, the 56-inch pipeline that runs 2,860 miles from Yamburg to the Soviet Union's western border is scheduled for completion in 1987. This line will deliver 20–22 Gcm/year of gas to other eastern European countries and the western part of the Soviet Union. It is supposed to have 40 compressor stations within the Soviet Union. Soviet gas production is one of the most successful activities in the Soviet economy, and according to the government of that country about 200 million Soviet citizens now have gas in their homes.

As noted above, the Soviet Union employs some pipelines with a diameter of 56 inches. This is the maximum diameter in use in the world at the present time, although some years ago it was believed that pipelines of 100 inches could be constructed. As things developed, however, it was decided that with the present technology 64 inches was the largest feasible diameter. Within this constraint research is concentrated on obtaining higher working pressures, and cooling the gas that is being transported. Multi-layer pipelines of 56 inches that are capable of carrying gas at a temperature of -30 degrees centigrade have been developed in the Soviet Union. Where length is concerned, the Siberian-Western European pipeline is about 4,500 kilometers at present, and may possibly be extended — which merely involves adding additional compressor stations. The longest pipeline however would be the one proposed to run from Northern Alaska, through Canada, to the 'lower 48'. Should this line be completed it would be at least 7,700 kilometers (= 4,800 miles) long, with one lateral running down the West Coast and another toward the Midwest. The well-known diagram in Figure 3.2 shows the relative cost of transporting gas by different means. It appears that the crossover point at present for LNG and onshore gas is about 6,500 kilometers, but clearly this type of deduction is inappropriate in many situations. For example, situations in which

Figure 3.2: Cost of transporting natural gas

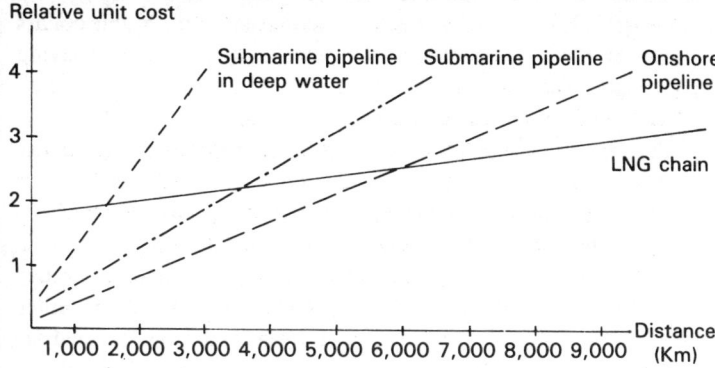

deliquefied LNG must be fed into onshore pipelines.

In 1979, the figures on the vertical axis represented the actual unit cost of natural gas in dollars/MBtu (Figure 3.2). With reference to the LNG chain the liquefaction cost at that time averaged 1.25 dollars/MBtu, while the regasification cost was 0.35 dollars/MBtu on average. As shown in the diagram, considerable expense can be associated with the construction of offshore pipelines, particularly those having to traverse deep water. One of the most impressive underwater lines now goes from the Hassi R'Mel field in Algeria, through Tunisia, and then underwater when it reaches the Mediterranean. It proceeds through the Sicily Channel and the Straits of Messina to Italy. The underwater crossing features the laying of packages of three 20-inch pipes at a depth that at times may reach 600 meters. Other underwater pipelines have been proposed between Algeria and Southern Europe (to France and Spain), but the technical problems presented by water depths of up to 2,000 meters probably cannot be solved at present within a suitable financial framework. Other notable underwater pipelines are in the North Sea, Caspian Sea, Gulf of Mexico and on the bottom of Lake Geneva — or Lake Leman as it is sometimes called.

LIQUEFIED NATURAL GAS

Now we turn to the matter of LNG, where one of the most important issues to be taken up below is obtaining the per-unit economic value of natural gas to exporters after deduction of all delivery expenses.

This is sometimes referred to as netback, and if desired it can be calculated at various stages in a conventional gas delivery system. First of all, however, the reader is presented with a general discussion of LNG, to include its background and the present world market situation.

The first LNG project commenced operation only in 1964, and in 1983 there were 16 LNG trading operations underway, delivering about six billion cubic feet per day (= 6 Gcf/d) of natural gas. This was four per cent of global gas consumption, and about 20 per cent of all gas traded internationally. Approximately 50 LNG tankers were engaged in this trade. Japan is the largest importer of LNG, and the other importers are France, Spain, Italy, Belgium and the United States. Algeria is the largest exporter, and Algeria, Indonesia and Brunei account for approximately 85 per cent of exports.

LNG projects must be based on comparatively large amounts of reserves: the capital costs of these projects are high, and for these projects to be profitable they must process and transport a great deal of gas. It is clearly suboptimal if the gas runs out while the facilities for liquefying and regasifying the gas have a decade or two of operational life left — unless, of course, the price of gas is high. Not only must adequate reserves be present, but buyers and sellers must be willing to commit themselves, via long term contracts running 15 to 25 years, to deliveries and acceptance of a certain amount of gas at prices that could be fairly rigidly established, although some variation in the price might be acceptable according to a formula that was previously accepted by both suppliers and consumers. It has also become usual to agree on an arrangement that allows price negotiations every few years.

As with pipeline gas, there are considerable economies of scale that can be obtained by increasing the size of LNG projects. It has also been suggested that projects involving less than 300 million cubic feet per day (= 300 Mcf/d) of gas are inherently uneconomical, although it may be true that projects in which the gas is transported a comparatively short distance, and serves several terminals, can be viable. However, on the basis of calculations that I have carried out, the assumption behind this belief must be large reserves in the country supplying the gas, and real prices for the delivered gas that are rigidly constrained against downward movements. Something that should be pointed out here is the importance of the discount rate in calculating the profitability of LNG projects. With very high investment costs, long periods of construction, and benefits (i.e. revenues) that accrue over long time horizons (up to 25 years or longer) profitability is sensitive to rising discount rates. The rise in discount rates over the past few years, which is associated with the rise in interest rates, has made many potential LNG projects unviable that were highly regarded five years ago. (The relation between discount rates and interest rates are not discussed here, but in certain ideal or textbook situations they are the same thing.) Table 3.1 shows both existing and proposed LNG projects as of 1 January 1984. For those that are proposed there is no specification as to who the eventual buyer will be. Many of the proposed projects have now been scrapped but they could be revived in the future.

There are 4 distinct phases in an LNG project: (i) gas production, treatment, and transportation to a liquefaction plant; (ii) liquefaction, storage and ship loading; (iii) shipping LNG in special cryogenic tankers to the reception terminal; and (iv) receiving and unloading the LNG, storing it and regasifying. These operations can be examined individually.

The gas that is used in an LNG project is essentially the same kind that flows into a pipeline. At the same time greater certainty must prevail about the quantity (and quality) of the gas, because for economic reasons the liquefaction plant should operate as close to full capacity as possible at all times. Prior to entering a liquefaction plant the gas should be carefully treated, and most impurities removed. If the gas contains a high percentage of carbon dioxide, hydrogen sulphide, nitrogen, and particles which could disrupt the production process, additional treatment is required.

Liquefaction involves cooling methane to about minus 161° Celsius and reducing its volume to 1/600 of gaseous methane. The cost of a 500 Mcf/d liquefaction plant in the United States, in 1983, varied between 900 and 1,100 million dollars. This is usually the most expensive link in the LNG chain, and the most expensive part of this activity are the steam and power generating facilities and the liquefying unit itself. It is generally believed that economies of scale (i.e. declining unit costs) prevail for liquefaction plants up to capacities of 1,500 Mcf/d, or perhaps even larger. But by way of contrast, unit costs rise rapidly for plants that are smaller than 500 Mcf/d. Liquefaction plants are capital intensive, and operating costs (= maintenance + labour costs) are only about five per cent of capital costs. When considering operating costs, the energy required to operate the liquefaction plant should be included. This averages roughly ten per cent of the gas input if gas is used as the source of energy.

Table 3.1: Operational LNG projects and projects under construction; proposed projects at 1 January 1984 (volume in Mcf/d; price in dollars/kcf ($\approx$ $/MBtu)

Exporter	Importer and scheduled initial delivery date	Years/volume	Price[a]
Algeria	UK, British Gas, 1964 (contract terminated)	15/110	—
Algeria	France, Gaz de France, 1965	25/50	5.870
Alaska	Japan, Tokyo Gas and Tokyo Electric Power, 1969	15/135	5.900
Libya	Italy, SNAM, 1970	20/240	—
Libya	Spain, INAGAS, 1970	15/110	—
Algeria	France, Gaz de France, 1972	25/350	5.87
Brunei	Japan, Osaka Gas and Tokyo Electric Power, 1972	20/745	5.93
Algeria	Spain, ENAGAS, 1976	23/450	—
Abu Dhabi	Japan, Tokyo Electric Power, 1977	20/355	6.22
Indonesia	Japan, Osaka Gas and Kansai, 1977	20/440	—
Indonesia	Japan, Chubu and Kyushu Power/Nippon Steel, 1978	20/630	—
Algeria	USA, Distrigaz, 1978 (small quantities since 1971)	20/120	—
Algeria	USA, El Paso, 1978 (suspended since 1980)	20/1000	—
Algeria	USA, Distrigaz, 1981	20/450	—
Algeria	Belgium, Distrigaz, 1982	20/500	6.10
Malaysia	Japan, Tokyo Gas, 1983	20/870	6.01
Indonesia	Japan, Nagoya-Osaka-Himeji, 1983	20/460	—
Indonesia	Japan, Niigata-Tokyo, 1983	20/480	—
Australia	Japan	20/850	—
Bangladesh	Japan/Europe	–/310	—
Cameroon	Europe	20/420	—
Canada	Japan/Europe (400 Mcf/d to Japan, 260 Mcf/d to Europe)	20/660	—
Indonesia	Korea/Japan (210 Mcf/d to Korea, 220 Mcf/d to Europe)	20/230	5.83
Nigeria	Europe	20/500	—
Qatar	Japan/Europe	20/870	—
Thailand	Japan	–/250	—
Trinidad-Tobago	USA	20/600	—
USSR	Japan	20/400	—

[a] information not available denoted by dash.
Source: World Bank documents.

When examining transportation, distance becomes important. If we continue to consider a project of the size given above, and assume that the LNG project is for North Africa-Europe, a fleet of five ships is needed to maintain an input at the consuming end of 500

Mcf/d. Each of these ships costs at least 150 million dollars. Operating costs generally constitute 7 to 8 per cent of capital costs if there is zero boil-off. The boil-off is the gas that is vaporised by heat leakage into the LNG tanks, and unless the ship is properly insulated — or has an efficient reliquefaction system — the boil-off can have a palpably unfavourable effect on the economies of the system. At the same time, though, the boil-off can be used to supply a portion of the fuel needs of a cryogenic tanker; and modern design has reduced boil-off to about 0.11 per cent/day.

Finally, at the terminal end of the chain a suitable harbour and regasification facility must be available. Specialised docking facilities for a 500 Mcf/d chain cost about 120 million dollars; the rest of the receiving terminal facilities averaged 350 million dollars in 1983. The total capital cost, on average, comes to 470 million dollars, and operating costs would have to be added in order to obtain total costs. Roughly, these operating costs are about three per cent of capital costs. In large installations the principal method for regasification was using sea water to raise the temperature of the LNG. Smaller installations (and peak shaving plants) use a process called gas vaporisation that features the employment of gas burners. The latter process is less capital intensive than the former, but lower capital costs tend to be offset by the cost of the large amount of gas that is used in the burners.

Calculation of netback

Figure 3.3 presents a schematic outline of an LNG project. At various points in this diagram are expressions for the calculation of netback, which has been defined above. One of the reasons for making this calculation at different points is because revenues from different activities often accrue to different persons. In the early period, before gas begins to flow, gross revenue is zero; in the periods when gas is flowing net revenue is $p_i q_i - E_i$, where p_i is the price of gas in period i, q_i is the quantity, and E_i are both operating *and* investment costs if investment costs are still being incurred. In the expression below, r is the discount factor which (as mentioned earlier in the exposition) need not be related to the market rate of interest, although this is very often the case. If the total netback accruing to the country delivering gas is greater than zero, or:

ECONOMIC THEORY AND NATURAL GAS

$$\overline{NB} = \sum_{i=0}^{i=T} \left[\frac{p_i q_1 - E_i}{(1+r)^i} \right] > 0$$

then the investment in an LNG scheme is judged more attractive than an investment in a safe financial asset. It should be appreciated that r might be chosen greater than the market rate of interest to account for uncertainty in the estimation of future costs and prices, although in theory uncertainty can be eliminated by long term contracts. Similarly, r might be less than the market rate of interest in order to take into consideration the social value of investment in LNG facilities to a country placing a high value on the employment and training provided by a project of this nature. Ideally, prices and costs are in constant values, and in the formula shown above the total netback $\overline{NB}$ is divided by the total amount of gas involved, V, in order to obtain netback in the conventional per-unit terms, or $NB = \overline{NB}/V$.

Figure 3.3: A typical LNG export chain

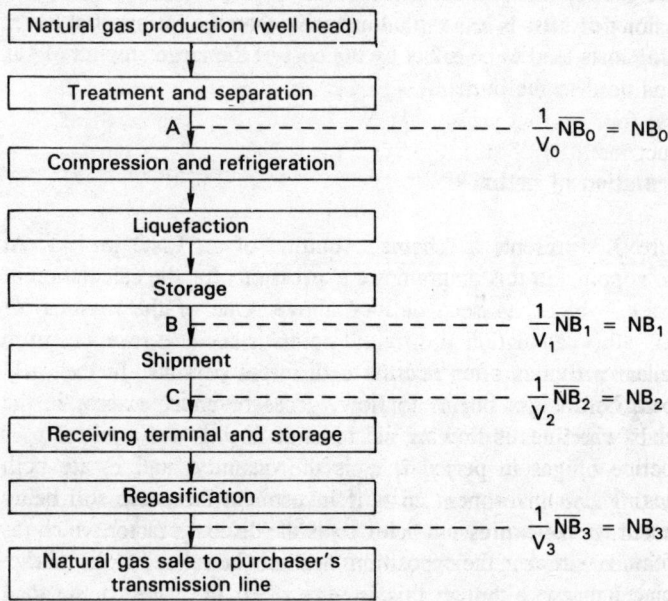

Note: A = gas sales at plant gate
B = LNG gas sales (FOB)
C = LNG gas sales (CIF)

52

There are generally three points at which netback is calculated. A (see Figure 3.3), the entry into the liquefaction plant (ex-pipeline); B, at the point where the liquefied gas is loaded onto LNG carriers (ex-liquefaction); and C, where delivery takes place to the importer (ex-ship). The last calculation in Figure 3.3, or $NB_3 = \overline{NB_3}/V_3$, gives the unit value of the gas as it enters the transmission pipeline of the importing country, and is thus simply the present value of the payments for gas, calculated at the regasification plant gate. Note, however, that in this calculation the payments $p_i q_i$ usually do not commence in the first year. Sometimes they do not begin until, for example, the start of the fifth year. If this is the case then, if $T = 25$, it means that the gas will flow for 20 years — from $i = 5$ to $i = 25$ — and usually during the first years of its flow it does not reach the maximum volume.

The first genuine netback calculation is ex-ship, or NB_2. Here we calculate the present value of net revenues throughout the time horizon of the project, making sure to include investment costs as well as variable costs. One of the reasons for including investment costs in this type of calculation is that such inconvenient concepts as 'interest during the period of construction' can be ignored. The expression for NB_2 is shown below, where E_{2i} is investment and operating costs for the receiving terminal, storage, and regasification facilities. Note also that investment costs largely terminate after the first four or five years, which is usually the time required to construct facilities of this type. Our formula is:

$$NB_2 = \frac{\overline{NB_2}}{V_2} = \frac{1}{V_2} \sum_{i=0}^{T} \left[\frac{p_i q_i - E_{2i}}{(1+r)^i} \right]$$

Similar calculations for netback ex-liquefaction add investment and operating costs of shipping natural gas to the above costs; while netback ex-pipeline requires the inclusion of all costs downstream of the pipeline delivering gas to the liquefaction (and pre-liquefaction) facilities. Thus, to the cost for netback ex-liquefaction are added investment and operating costs for liquefaction and storage.

Before a simple numerical example is given it should be appreciated that in a perfect system no gas would be lost between the entry point to the liquefaction system in the gas-producing country and the entry to the transmission pipeline in the gas-purchasing country, but these systems are not usually perfect.

Among other things there is the boil-off of gas in ocean-going tankers that was referred to earlier. In many calculations of netback, however, it is assumed that there are no losses or negligible losses in the LNG chain, and thus in the above calculations we would have $V_0 = V_1 = V_2 = V_3$.

Now for a simple example that illustrates the concepts that have been presented in this section. Suppose that a quantity q of 2,500 units of natural gas is to be delivered to a buyer. The delivery dates are as follows: 500 units during the third year of a five-year project; and 1,000 units in the fourth and fifth year of the project. Payment is at the beginning of the year, and the price for the gas in each delivery year is 10 dollars/unit. The investments that are required for this project cost 4,000 dollars at the beginning of the first year, 6,000 dollars at the beginning of the second year, and 2,000 dollars at the beginning of the third year. Operations and maintenance costs are 1,000 dollars/year beginning in the third year. The discount rate is taken as ten per cent. Schematically, this appears as follows:

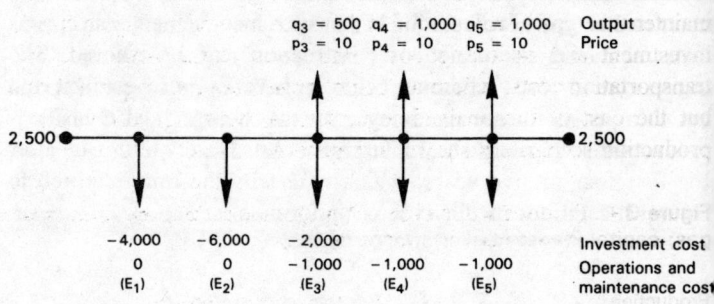

Employing formulae of the type presented above we have:

$$NB = \frac{1}{2,500}\left[\frac{0-4,000}{1.0} + \frac{0-6,000}{1.1} + \frac{5,000-3,000}{(1.1)^2} + \frac{10,000-1,000}{(1.1)^3} + \frac{10,000-1,000}{(1.1)^4}\right] = 2.042 \text{ \$/unit}$$

A positive netback indicates that if cost and benefits (i.e. revenues) are as indicated above, this particular physical investment is — from the point of view of profit maximisation — more satisfactory than an investment in a safe financial asset (such as a government bond). In other words, an economic profit can be obtained if this investment is made.

This discussion has ignored fiscal and financial cost components (for example, depreciation, depletion, taxes and subsidies). In actual calculations this often makes sense, particularly when the gas is being produced by a public company, and in the importing country taxes are the concern of the transmission company and similar agents.

OFFSHORE GAS

To conclude, I would like to make a few observations about offshore gas fields. Gas from these fields can provide the input for both pipeline gas and LNG. As with the previous topic, the delineation of costs is of importance when considering offshore operations. Costs are again divided into investment costs and operating costs, where investment costs are platforms, machinery, pipelines, and the cost of putting equipment and structures into place. Operating or variable costs are as usual wages, the cost of material inputs, maintenance costs, and so on. In Norway it is customary to divide investment and operating costs into costs at the wellhead, and transportation costs, where the latter include not only pipeline costs but the cost of terminals, buoys, etc. A typical investment and production scenario is shown in Figure 3.4.

Figure 3.4: Production/investment profiles (production in units of gas; capital investment in money terms)

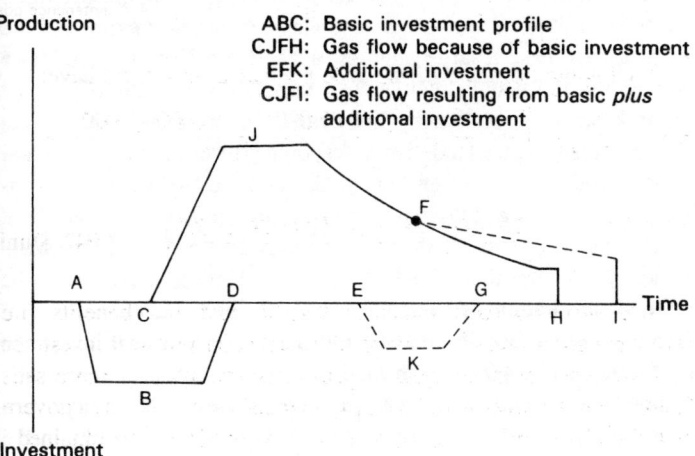

ABC: Basic investment profile
CJFH: Gas flow because of basic investment
EFK: Additional investment
CJFI: Gas flow resulting from basic *plus* additional investment

It is clear from Figure 3.4 that the production profile can be modified by changing the investment profile: less investment means lower peak production or a shorter period over which the gas flows, or both. The reader might wonder about the significance of the investment profile EKG, since it seems to be detached from the main body of investment ABD. EKG could represent investment that takes place after the discovery of additional reserves, and leads to a modification of the production profile from CJFH to CJFI. It also turns out to be the case that peak production is generally a function of the total reserves that are available. In the North Sea, for example, peak production in a surprising number of cases appears to be somewhere in the neighbourhood of nine per cent of the reserves in a gas field.

The cost of obtaining offshore gas is a direct function of the water depth and the number of production wells — where the latter is determined by the desired output. Economists often attempt to complicate the problem of determining the investment profile which, in my view, is fairly simple in practice. A certain amount of gas exists for which the seller desires to find a market. At the same time there is (hopefully) a range of possible buyers for the gas, who are prepared to back their interest in the gas with enough financial resources to make exploitation profitable. Negotiations between seller and buyer follow, and a production profile is eventually determined that is satisfactory to both parties. (Behind the production profile, of course, is an investment profile that makes it possible for the producer to supply the desired output, and at the same time make at least a 'normal' profit.) For instance, in one of the major proposed gas deals of the century, the British Gas Corporation expressed the desire to purchase a large amount of gas from Norway's offshore Sleipner gas field; and although the transaction was eventually called off because of objections by the British government, a price was agreed upon as well as a delivery schedule. Here it can be noted that in the negotiations between British Gas and the Norwegian government, as well as the subsequent decision by the British government to veto the arrangement, there were many non-economic or quasi-economic factors that weighed heavily. Among these were the political orientation of potential suppliers, and the very different perceptions that British Gas and the British government had about the future availability of domestic gas.

One of the main issues that inevitably arises in those countries where the exploitation of offshore deposits is not completely in the hands of governments concerns the bidding for offshore leases —

where these leases give the holder the right to explore for and produce gas in a specific block of offshore territory. The principal issue here involves depriving private companies of the super-profits that can result in the event of a highly successful operation. The means chosen to do this is competitive bidding which, if a large enough number of firms are participating, will result in firms with superior knowledge offering such high prices for offshore leases that, on the average, only normal or subnormal profits will accrue to firms that obtain these leases. A number of investigators have claimed that of the more than 1,000 oil and gas leases issued by the United States government to firms operating in the Gulf of Mexico in the 20 years between 1950 and 1970, only a few resulted in an excessive return on investment. In fact it has been said that the government received such high payment, and firms such low profits, that the offshore sector comes very close to being a competitive market on the producing side.

It also appears that the joint ventures often resorted to in the offshore sector encourage competition, in that they make the participation of smaller firms in offshore drilling more likely — but as far as I am concerned it would be too much to expect that any method of dispensing leases would allow truly small oil and gas companies to compete with the large oil firms in situations where these large firms were truly determined to secure exploitation rights.

The exact design of a lease cannot be dealt with here, but it is clear that leases can be very powerful vehicles for relieving their purchasers of excess cash. Sometimes there is a stipulation in the lease that if oil or gas is discovered and marketed, the seller of the lease must obtain a share of the profits. Often the purchase of a lease commits the buyer to royalty or rent payments. As yet it appears that no foolproof formula exists for selecting the perfect lease, nor does it appear that economic theorists have constructed any realistic models that the buyers or sellers of leases can use to determine an optimal course of action in real world situations.

4

Natural Gas in the United States

This chapter will consider natural gas in the United States, although a few observations will be made about the situation in Canada and Mexico.

It has been clear for a long time that the United States has the largest and most complex natural gas sector in the world. In 1982 there were almost 4,000 corporations producing natural gas in the United States, 63 inter-state gas transmission companies — several of which were among the largest corporations in the country, and which in addition produced natural gas and/or oil — and approximately 1,700 distribution companies. These enterprises had at their disposal 78,000 miles of gas-gathering lines, 263,000 miles of transmission lines and about 688,000 miles of distribution mains. The natural gas sector is mostly private, with the exception of about 500 public utilities that supply private consumers.

Very large pipelines were laid from Texas to the Midwest, Pacific Coast, and the North East during and after the Second World War. Gas was bought for 5 to 10 cents per thousand cubic feet (= 5–10 cents/kcf) in Texas or other parts of the South West and made available in New Jersey or New York for 25–30 cents/kcf. The buyers were usually utilities who piped it under metropolitan streets to final consumers. In 1975 the average price of gas to a consumer in New York was 1.88 dollars/kcf, which is approximately equal to 1.88 dollars per million Btu (= 1.88 dollars/MBtu). Of this amount 20 cents went to the producer in the field, 22 cents to the pipeline transporting the gas, and the rest to the utility distributing the gas.

Despite its many achievements, the natural gas industry in the United States is experiencing a special sort of anguish. First of all, it is faced with a temporary oversupply of gas which has been termed a gas bubble. This gas bubble, which is estimated to be between 1.5

and 2.5 Tcf, has probably helped to hold down the price of gas, but has not convinced gas buyers that they should sign new long term contracts at what the gas industry regards as reasonable prices. As pointed out below, many buyers are waiting to see what is going to happen with coal and oil, and in addition there have been many predictions of a declining availability of gas, although some observers disagree and predict surplus delivery capacity until 1995. Some buyers are also worried about the deregulation of gas prices that has been taking place over the past few years, although as yet gas producers have not been able to take advantage of this phenomenon to the extent that they desire because of the gas bubble. In 1985 United States gas production should have been about 18 Tcf. An average of forecasts indicates that about 17 Tcf will be produced in 1990, and only 13 Tcf in the year 2000. There has been a tendency to increase the amount of spot sales of gas, where the spot trading of comparatively small amounts of gas (in individual blocks) can be contrasted with trading large amounts, over many years, in a more or less continuous stream, via the medium of long term contracts.

SUPPLY, PRICE AND DEMAND OF NATURAL GAS

In examining the supply of natural gas in the United States, we are immediately able to say something about the price. The stock of gas in the ground is less accessible today that it was five years ago, and very likely more accessible today than it will be in the future. What is happening is that the productivity of individual gas wells is progressively decreasing: less gas is being obtained per well, and drillers must go deeper after gas. It seems almost certain that where conventional natural gas is concerned, this trend will not be reversed. *Ceteris paribus*, this will tend to push up the price.

By the same token, drilling costs are increasing. This is a much more difficult topic to handle because drilling costs can change for other than geological reasons. For instance, they can increase if there is full employment of drilling crews and equipment (as was the case between 1980 and 1982). Aside from this, it appears that between 1960 and 1974, drilling costs increased by about two per cent a year, and some acceleration in costs may have taken place after 1984. In 1960 drilling costs (in current dollars) were approximately 47 dollars/foot. This increased to 60 dollars/foot in 1974, and approximately 130 dollars/foot in 1982. Expectations are that this cost could go to 180 dollars/foot by the turn of the century, which should also tend to increase the price of gas.

The average wellhead price of natural gas in the United States increased from about 0.50 dollars/Mcf in 1973 to 1.18 dollars/Mcf in 1979, to 2.50 dollars/Mcf in 1982. As will be discussed below, widespread deregulation of gas prices is taking place in the United States, but even 2.50 dollars/Mcf was still about 40 per cent below the average world market price during 1982. In the middle of 1985 the spot price of gas in the United States was about 2.75 dollars/Mcf, and expected to fall. Certainly, no one expects this price to reach the 3.73 dollars/Mcf (on the average) that is presently allowed by federal regulations — at least not in the near future. In negotiations dealing with natural gas that is being purchased under long term contracts, the base price has apparently been in the neighbourhood of 3-to-3.50 dollars/Mcf; but as mentioned earlier this price is usually indexed in some way, or long term contracts are drawn up in such a way as to permit its renegotiation on a periodic basis or if there is drastic change in market conditions. It is exactly this provision that many buyers fear.

Before turning to demand we can look at another important factor determining both the supply of and demand for natural gas. Because of the extremely high investment costs that characterise various components of a natural gas project, it is imperative that large reserves are available. Otherwise it is difficult to justify pipeline systems and, on the demand side, equipment in which the gas is used. In addition, reserves influence drilling costs, because the more gas that is available in an area, the easier — and less expensive — it will be to find and exploit it.

In 1983 the 30 largest United States energy companies found only enough gas to replace two thirds of the amount they sold. The simple fact of the matter is that if the demand for gas increases or is constant, the reserves of conventional gas in the United States must fall. The production of gas from a gas reservoir is dependent on the reserves in that reservoir, because (as with oil) output should not rise above a certain fraction of reserves. With oil this figure is about one-ninth or one-tenth of reserves, while with gas in the United States one-seventh is the figure that is often quoted (put another way, the critical reserve/production ratio is seven). If the reserve/production ratio becomes lower than the critical ratio, some of the gas reservoir is destroyed (see *The Political Economy of Oil*). The reserve/production (R/Q) ratio was ten in 1960, and fell to less than nine in the late 1970s and early 1980s. It rose above ten in 1983 when demand fell sharply. Some observers believe that it will fall below ten again in the future if suppliers have their way, because gas producers — as a group — seem to feel that even if they can sell large amounts of gas, they will succeed in finding enough to prevent R/Q ratios from dropping beneath the critical level. It also seems to be the case that if probable reserves are taken into consideration, the R/Q ratio in the United States is about 20 at the present time.

On the demand side gas has industrial uses, is used to generate electrical power, and is an important energy source for households. About 55 per cent of all United States households (or about 42 million households) use natural gas for space heating, and another seven million use gas for other purposes. Conservation, however, is leading to falling demand for gas by households, and the shrinkage of such energy intensive industries as steel, cars, and metal smelting and refining is resulting in at least a temporary loss of these important sectors to the gas industry. In addition, as pointed out below, gas may be rapidly losing ground to coal where electrical power and industrial uses are concerned, and co-generation has become an important source of industrial energy. As an example, General Motors, which has been a major purchaser of natural gas supplies for about three decades, now uses almost 30 per cent less natural gas to make a vehicle than it did 12 years ago. This lower energy requirement can be regarded as permanent, although higher industrial production in the United States in 1984 caused industrial gas consumption to be at least 15 per cent higher than in 1983.

Gas industry spokesmen often suggest that their colleagues will have to counter the present malaise in gas demand by more aggressive marketing, and it is also maintained that a more flexible price policy must be established, with more competition between gas producers and pipelines. What all this comes down to is more spot sales, and more direct sales between gas producers and large industrial consumers that bypass pipeline companies. In other words, the producers construct pipelines and/or make arrangements with pipeline companies to reduce supply when industrial consumers require less gas. Analogously, some pipeline firms should move more deeply into production. In particular, scope must be found for consumers to avoid having to take (or pay for) gas that they do not need, because in the long run these consumers shift their loyalty to another energy medium.

A determining factor in the demand for natural gas appears to be the amount of gas using durable equipment that consumers have installed and the length of life of this equipment. As it becomes obsolete, or loses its operating efficiency, decisions have to be taken as to whether it should be replaced. In general, if the price of gas is relatively low in relation to other energy sources, and it appears that this will be the case over a large part of the life of the gas-using equipment, then homeowners, industrial firms and utilities favour this kind of equipment over its competitors. As things now stand in the United States, with gas apparently plentiful and a downward pressure on gas prices, many consumers who are already using gas and are in a position to do so are renewing their commitment to gas. For the United States suppliers of gas this is good news, but they are not particularly happy about the possibility of a larger share of this non-expanding market being taken by imports of either pipeline gas from Canada or Mexico, or LNG from North Africa.

Thus far electricity has not been mentioned, but gas and electricity are often direct competitors in many household and industrial applications. Gas is probably holding its own in the United States since the supply of nuclear-based electricity has lost its momentum. If the situation in that country is similar to what is happening in Sweden, many potential consumers of nuclear power are now thinking in terms of gas. The situation with oil is more complicated. In the long run gas should easily succeed in maintaining its position *vis-à-vis* oil for the simple reason that there is more of it available both locally and in the world outside the United States. But in the short run some not so obvious changes in demand and price can come about. If the price of oil falls by 10-to-20 per cent, then both the demand for gas and its price should increase. This is because a fall in the price of oil of this magnitude should stimulate international economic activity, and increase the demand for all energy resources and their price. In the longer run, if the price of oil were to fall by more than the amount suggested above, then a great many consumers of gas and coal would begin thinking of using oil. In fact, where installations have dual capabilities (as discussed below), the shift from gas to oil would take place immediately, and some consumers might also lock themselves into an extended relationship with oil by signing long term contracts with suppliers who are afraid that the price of oil will fall to the four-to-eight dollar/barrel range suggested by some analysts. Within the upper managerial echelons of the petroleum industry, it is generally believed that a sizable fall in the price of oil would only lead to a more striking rebound in the price later on.

Just now natural gas is winning the battle against oil in United States bulk fuel markets but is being outpaced by coal. In 1973, of all bulk fuels consumed in United States industrial and electric utility activities, coal held a 25 per cent share, as compared with 28 per cent for gas and 25 per cent for oil. At the beginning of 1985 these figures had changed to 31 per cent, 20 per cent, and 17 per cent. (The two main bulk fuel markets in the United States are electrical utilities and the industrial market. In 1984 coal's share of the utility sector was 54 per cent as compared to twelve per cent for gas and five per cent for oil. In the industrial sector the share of coal was ten per cent, while that of gas was 26 per cent and oil's share was 29 per cent.) The reason has been the relatively low price of coal, particularly since 1979 or 1980, when there was an increase in the supply of coal due to a fall in the demand for United States coal in foreign markets. In addition, the passage of the Natural Gas Policy Act in 1978, which deregulated the price of some gas, was interpreted by many energy purchasers as meaning that all gas prices would soon be free. These buyers tended to disregard the claims of gas suppliers that the removal of federal control of wellhead gas prices would eventually provide suppliers with the incentive to find enough gas to keep prices from exploding upwards.

The consumption pattern in the bulk fuel markets cited above should prevail in the long term, with gas remaining strong in the industrial market and coal dominating the power generating market. The price advantage held by coal over gas will almost certainly remain undisturbed, and this will prove decisive in those activities where simple steam raising is the objective — such as power generation; but as a process furnace fuel, coal is often unsatisfactory. A coal combustion chamber requires three or four times the space of an oil or gas furnace, and as a result there are major problems associated with controlling the temperature: to decrease the temperature in an oil or gas furnace, the fuel rate is simply decreased, but in a coal-fired furnace it is usually the case that the temperature cannot be lowered before the contents of the furnace have been burned. Imaginative planning may ensure that on those occasions when lower temperatures are likely to be desired, the furnace does not contain a great deal of coal, but clearly the flexibility of this type of installation is greatly reduced in that considerable demands are placed on what might be scarce planning capacity.

Something that has tended to increase the competitiveness of gas (and oil) in general has been the development of dual fuel use capability. What this means is that one installation can burn gas or oil. Between ten and 15 per cent of the electrical power generating capacity on the East Coast has dual capabilities, and according to the American Gas Association a huge amount of gas sales (equivalent to 510 Gcf/y) were lost during the first quarter of 1983 due to the increased use of oil by these installations. Gas is also making impressive progress in the co-generation market, where the object is to harness the waste heat that otherwise escapes from an engine or turbine and put it to industrial use. It appears that gas is the main choice as a fuel in co-generation markets.

Finally, coal has been given an important advantage in the United States by the Fuel Use Act, which requires that new boilers use non-gas, non-oil fuels. In the short run this may not have a great impact, because existing boilers do not have to be converted, but in the long run, oil and gas usage in boiler fuel markets must decrease drastically.

SPOT MARKET FOR NATURAL GAS

Now let us devote some attention to the matter of the spot market. The tradition in natural gas has been, as explained above, to sell gas on the basis of long term contracts: neither buyer nor seller can afford to have something go wrong when millions — or hundreds of millions — of dollars have gone into expensive investments in physical capital. But in the presence of the gas bubble, and the growing tendency of the legal system (i.e. the courts) to support gas purchasers who wish to set in abeyance the take-or-pay contracts under which they bought gas, a large market for spot gas has come into existence. Many distributors of natural gas are now in position to go out and buy it in the same manner that Coca-Cola and ice cream are purchased, except that delivery can be delayed for up to six months. In addition to distributors, the spot market is now being used by electric utilities, chemical producers and similar enterprises which previously bought all or most of their supplies on long term contracts. Now, to an increasing extent, they are turning to spot market brokers who are in touch with producers who are willing to sell at lower prices.

In 1984 spot gas sales came to about six billion dollars; preliminary estimates for 1987 suggest that they could reach ten billion. The spot market has developed into a price discovery mechanism for the natural gas industry as a whole, and an instrument for bringing about equilibrium in markets where long term contracts are signed without future demand being known. Table 4.1 gives listings of

NATURAL GAS IN THE UNITED STATES

Table 4.1: Spot market prices in the United States (March 1985)

Pipeline system	Point of receipt	Price (dollars/MBtu)
Columbia Gas	Louisiana Laterals	2.95
El Paso Natural Gas	Pecos and Waha, Texas	2.65
Houston Pipe Line	System Wide	2.80
Panhandle Eastern	Beaver County, Oklahoma	2.70
Tennessee Gas	Vinton, Louisiana	2.80
Texas Eastern	Zone A	2.80
Texas Gas Transmission	North Louisiana	2.85
Transcontinental	Production Area	2.90
ANR Pipeline	Beaver County, Oklahoma	2.60
Colorado Interstate Gas	Beaver County, Oklahoma	2.60
Natural Gas Pipeline	Beaver County, Oklahoma	2.60
Northern Natural Gas	Beaver County, Oklahoma	2.60
ANR Pipeline	Garden City, Louisiana	2.90
Louisiana Intrastate Gas	Garden City, Louisiana	2.90
United Gas Pipeline	Garden City, Louisiana	2.90

Source: United States Natural Gas Clearinghouse (April 1985).

regional United States spot market prices for March 1985.

The prices in Table 4.1 were reported by the United States Natural Gas Clearinghouse. This is a privately owned corporation formed in 1984 to provide in the words of its owners, 'one-stop shopping for users of the gas spot market', and its first transaction was in October 1984. By March 1985, transactions totalling 22 billion cubic feet of gas had been handled by the clearing house, and these involved about 300 end users, 19 pipeline systems, and 150 producers. Most of the transactions took place in the Rocky Mountain and Midcontinent region, and to a certain extent the Northeast. For the most part the clearing house is dealing with a large number of brokers who represent smaller independents, and apparently one of the things making this arrangement work so well is that the producers taking part are already associated with pipeline companies. Equity partners in the clearing house are Morgan Stanley, Colorado Interstate, Columbia Gas, El Paso Natural Gas, Houston Natural Gas Corporation, Transcontinental, United Gas Pipeline, and the Washington DC law firm of Akin, Gump, Strauss, Hauer and Feld.

It has been claimed by some observers that in the wake of large scale gas price deregulation, the spot markets can capture as much as 20 per cent of the United States gas market in five years. As far as I am concerned, that figure is too large, because in order for that much gas to move in spot markets, gas brokers would have to have access, at all times, to a nationwide grid of gas pipelines: consumers will not enter into arrangements to buy gas unless it can be delivered to them when they want it delivered. Pipelines with idle capacity will, of course, be willing to transport spot market gas, and the United States Clearinghouse system claims that more than 100,000 miles of gathering and transmission lines of its associates has access to at least 90 per cent of the United States gas market; but even so this is going to be a herculean task. It is here, again, that we see one of the great advantages of oil, which can be transported by a variety of media (pipelines, ship, tanker, truck, etc.) and easily moved from one mode of transport to the other.

REGULATION AND DEREGULATION

In examining the literature on natural gas in the United States, it seems quite clear that regulation is one of the major topics.

When so much is written against the regulation of business by government, some reminder should be given as to why the gas market became so heavily regulated. To begin with, there were safety factors. Natural gas is dangerous unless it is properly processed and constrained, and even now there are occasional fatal accidents caused by explosions of natural gas. Earlier in this century, however, there were hundreds of these accidents; and when the regulatory function was the responsibility of individual states, those states in which regulations were not introduced or enforced presented a markedly higher casualty count than those in which the sellers of natural gas were held responsible for defects in the gas bought by households and others.

But there is also an important economic consideration that has already been mentioned but needs to be stressed even more. Because of the huge investments that are involved, the consumers of natural gas generally have access to only a few suppliers. The model of perfect competition that we are familiar with from the economics textbooks is irrelevant. If households, and others, have invested in expensive equipment so that they can use natural gas, they are in a difficult situation if the price of gas unexpectedly increases: where heating and cooking are concerned, many households have little flexibility, and as a rule do not have access to alternative suppliers. It was for this reason that some of the sharpest regulatory legislation in the history of the United States was passed during the presidency of Dwight D. Eisenhower, perhaps the most conservative United

States president of the post-war era. When tested in the Supreme Court of the United States, these regulations were deemed suitable by such jurists as Sherman Minton, perhaps one of the most conservative Supreme Court judges of this century. (The case in question was the 1954 Phillips decision in which the court, invoking the General Welfare Clause, ruled that the Federal Power Commission — which is now the Federal Energy Regulatory Commission — would control the price of most gas flowing in interstate commerce. *Intrastate* gas sales were not affected by this decision.)

There is no reason why the price of gas should not increase unexpectedly. Although in theoretical work it is often convenient to assume perfect foresight on the part of investors and consumers, in reality most of us have only a vague idea of what the price of essential consumer goods will be in a year's time, and no idea of what can happen in five years, or a decade, or longer. The reasoning that was employed by the people who framed regulatory legislation for the natural gas market 30 years ago was that, if the price of gas had climbed suddenly by a large amount, it would have been economically disastrous for millions of low and middle-income households which were locked into gas, and had counted on low gas prices when they bought gas stoves and heaters.

Distributors and households were favoured, but pipeline owners and producers were disadvantaged because the focus of regulation was at the wellhead. Since the deregulation has begun, many producers feel that they are better off, but consumers complain that they are being mistreated, and many pipeline companies are also dissatisfied.

The difficulty begins when prices at the wellhead increase, and thus pipelines are faced with costs that consumers (= distributors + end users of gas) are unwilling to absorb. Producers and pipelines are linked by long term take-or-pay contracts, of varying severity, by which it is meant that buyers must take the gas contracted or pay for all or a part of it. Similarly, contracts between pipelines and distributors, which are sometimes called city gate contracts, are also long term, and usually contain minimum take or minimum bill clauses which stipulate that a distributor must purchase at least a minimum amount of gas from a pipeline company at prices based on the pipeline company's costs. Contractual arrangements of this type are intended to spread risks over all parties, since they secure the availability of gas to distributors, while helping to reduce the losses that the most capital intensive part of the natural gas chain, the pipeline, might suffer from not being able to operate at intended capacity. They also increase the ability of the pipeline companies to meet take-or-pay obligations at the wellhead.

City gate contracts also tend to contain sole supplier clauses, which mean that a distributor can purchase gas only from a single pipeline company; and territorial restriction clauses, which restrict the geographic area in which a distributor can resell gas purchased from a particular pipeline company. Clearly, the ability of distributors to recover costs is determined not just by the willingness of regulators to change the rate structure, but also the elasticity of demand for gas at the burner tip, since the distributor does not (and cannot) have a contract with his customers of the type he has with the pipeline company. Industrial customers with dual capabilities have a highly elastic demand (in that rate increases cause them to go over to another fuel), while it is very often the case that higher prices will not result in residential consumers lowering their consumption of gas by a large amount.

The reason for the rapid deregulation of gas — in the face of the excess supply that exists at present — is the excess demand that prevailed until recently: gas was being demanded at a much higher rate than producers were able to provide it, and gas was being consumed at a faster rate than new reserves were being discovered. The longer the price of gas stayed depressed, the more convinced many potential users of gas became that it made sense to install durable gas-using equipment, although for many of these investments to be considered justifiable, inexpensive gas would have to be available over much of the equipment life, and there could be no guarantee that this was going to be the case. What finally happened was that at existing prices so much gas was demanded that a system of curtailments and restricted access — which amounted to *de facto* rationing — had to be introduced, and this system gave some consumers, depending upon where they lived, as much gas as they wanted at low prices, while other consumers had to pay high prices for gas or do without.

Eventually, with a permanent gas shortage looming, and everyone complaining about the inequities of the arrangement cited above, it was deemed that drastic medicine was justifiable, and steps had to be taken to bring this complex market into equilibrium. This meant that raising the price of gas was unavoidable.

The Natural Gas Policy Act of 1978 (NGPA) was intended not only to bring the market into equilibrium, but to do it in an essentially painless manner. The price of gas would be allowed to increase, but in such a way as to promote increased supply, while minimising the discomforts that these price increases would impose on residential (and to a certain extent on commercial) consumers. The exact functioning of the NGPA was to be as follows:

> Federal price controls were to be extended to intrastate gas, and thus there would be a unified national system of wellhead price regulations.
> Natural gas would be classified into 22 different categories, depending on age, location, depth and production cost. Price ceilings would be established for all categories.
> New gas from deep or non-traditional sources (e.g. frontier gas) would be decontrolled immediately.
> Periodic price increases would be allowed for gas discovered after 1976, and the price of this gas would be completely decontrolled by 1985.
> Old gas, or gas discovered prior to 1976, would continue to be controlled; but provision was made for periodic price increases.

The average price level of gas, which took into consideration all categories of gas, rose much more rapidly than expected, especially after the price of oil escalated in 1979–80, and an appreciable fraction of the demand for energy that would have applied to oil devolved on gas. Recently, with the price of oil falling, and a large part of the world in recession, there is a downward pressure on natural gas prices. But the consensus is that the situation will not persist. In the long run reserves will probably not be found that will support the present rate of consumption, and unless prices rise, the greater (economic) evil of rationing must be resorted to. Another problem is that the price increases that have already taken place have succeeded in palpably lowering the standard of living of poor and lower income families in cold climates, since most of these families have no way of decreasing their consumption of gas, or switching to other sources of energy. The solution to this problem is almost certainly subsidies to low income families, but in a period of lower taxes and higher defence expenditure, it is probably impossible to initiate and/or implement this type of arrangement. Unless I am mistaken, the regulation-deregulation issue is one for which there will never be clear-cut answers, but an interesting attempt to organise some of the components of this issue has been made by Hubbard and Weiner (1984). The United States natural gas market is difficult to analyse because of the 'separateness' of producers, pipeline companies and distributors, who operate in a situation where future demand is important but imperfectly known. If winters are warm, households will use less gas for heating; and if the price of oil falls, industrial fuel purchasers with dual capacity will switch from gas to residual fuel oil. But when the final demand for gas falls, producers and pipeline companies who have invested hundreds of millions of dollars in placing gas at the disposal of households and industries will be in trouble, which is why sophisticated investors try to avoid putting their money into enterprises of this type unless provision is made to protect them from the vagaries of demand.

Another item in the United States regulatory jungle is the carrier status of natural gas pipelines. Pipelines can be classified as contract carriers which are paid to transport gas owned by other parties; or they are private carriers which move their own property. With this latter arrangement pipelines buy from producers, and then sell to large industrial consumers, electric power plants, or local distribution companies (city gate sales). But it has been suggested that since pipelines are in position to suppress competition by refusing access, they should be classified as common carriers. This would mean that it is mandatory for a pipeline to provide transportation services for a regulated fee. Accordingly, the pipeline system as a whole would become more flexible and responsive, and thereby competitive in the long run, which in turn could result in considerable benefits for both producers and consumers, as well as potential producers and consumers. (Potential producers in this context are producers who are operating well below their capacity, while potential consumers include gas users who have switched to an alternative fuel, but still possess gas-burning facilities. Clearly, there may be prices at which transactions can be carried out between these two groups, and if they could be brought together it would represent an unambiguous welfare gain.)

It seems relatively certain that many aspects of the complicated system discussed in this section will be phased out, and probably sooner than later. If pipelines become subject to the common carrier stipulation, which would almost certainly lead to a greater supply of gas at the city gates, then the minimum bill and sole supplier provisions of many city gate contracts are illogical, in that they would prevent distributors from changing from high-cost gas to low-cost gas, and also from selling their excess gas at the highest prices. Similarly, producers able to collect high prices from pipelines via take-or-pay contracts would be hesitant — in a highly flexible system featuring increased access — to sell at the low prices offered by some distributors because this low price gas might be resold.

Perhaps the best way for deregulation to evolve would have been through removing as many restrictions as possible at the wellhead and on the activities of distributors *vis-à-vis* pipelines, while making provision for pipeline companies to recover their costs and, at the same time, encouraging these firms to enter into more creative arrangements with producers and distributors. This means that pipeline companies must be permitted to vary their tariffs depending on the elasticities of demand on the selling end, and their ability to negotiate satisfactory prices with the producers of natural gas.

CANADA AND MEXICO

At present Canada has the capability to export 28–29 Gcm/y of pipeline gas to the United States, although actual exports are probably less. The same capabilities will remain at least through 1995. Mexico's export capacity to the United States is 3.1 Gcm/y, and this may rise to 10 Gcm/y by 1995.

Like the United States, Canada could be described as energy-rich. The largest reserves of natural gas can be found in Alberta, but there are also sizeable reserves in British Columbia, the MacKenzie Delta-Beaufort Sea area and the Arctic Islands. Gas resources in these latter regions are promising but will be costly to obtain. It is the intention of the federal government's National Energy Programme (NEP) to ensure that financial and fiscal incentives are available to stimulate the exploitation of frontier areas. The resources of these outlying districts are now managed and controlled to a considerable extent by the federal government, which is not to the taste of the provincial governments. It is also believed that Canadian reserves would be greatly increased if the so-called deep basin areas of Alberta and British Columbia could be exploited, but in order to do this important technological developments are necessary. As with unconventional gas in the United States, the appearance of this technology is taking much longer than expected a few years ago.

Mexico is also a country with large gas resources. Its current policy, however, is to concentrate on the export of oil, while maximising the local use of gas. One reason for this is an unpleasant experience Mexico had with the United States. In 1977 an agreement was anticipated by PEMEX (the Mexican state oil and gas company) that would involve Mexico selling 2 Gcf/d of gas to the United States; a letter of intent was signed by PEMEX and a consortium of United States gas companies. This deal was cancelled by the United States government. Later another deal was negotiated between the Mexican government and Border Gas Inc for a smaller amount of gas: about 300 Mcf/d at a base price of 3.625 dollars/MBtu. (As indicated below the price, when the gas started to flow, was higher because of the indexing formula; but Border Gas reduced its purchases.) At approximately the same time, however, a transaction was concluded with Canada in which United States buyers paid 4.47 dollars/MBtu for gas. The Mexicans interpreted these events in a negative light, and apparently came to the conclusion that they would be better off employing their natural gas in domestic activities whenever possible.

Canada is even better endowed with natural gas than such well known gas producers as Norway, The Netherlands and Britain (Table 1.2). The consumption of natural gas is also high in Canada, representing 22 to 23 per cent of total energy use, and expected to reach 25 to 30 per cent by 1990. The region with the largest energy demand is the eastern part of the country where the main energy resources are hydropower and coal. That part of the country, because of its relatively high population density, has the strongest voice in the federal government, and actively works for the federal government to have a greater influence in the disposition of the resources of frontier areas, particularly those in the West.

There is a strong feeling in Canada that energy prices should be kept low, which means that with gas prices rising in the United States, gas should not be exported if it means depriving Canadian consumers of supplies. At the same time there is a desire on the part of authorities and energy companies in Canada to take advantage of the higher United States prices, and recently this has made more gas available for export. There is no point in attempting to deal with the economic theory of this issue, because what Canada does or does not export is settled by politics and not economics. (The Canadian Energy Board has decided that Canada's exportable surplus of gas is about 500 Gcm, and of this the United States is eligible for about 325 Gcm. The price at which this gas may be exported has also been adjusted downwards as gas prices have softened in the United States.) There are several schemes being considered in Canada for the sale of LNG to Europe and Japan, but as far as I am concerned, given the present development of the world natural gas supply-demand situation, the schemes are non-starters. (For greater detail, see the later chapters.)

Mexico — like Canada — has, or at least had, high hopes of

exporting large amounts of natural gas to the United States. Several years ago, however, the organisation formed to buy gas from Mexico (Border Gas) reduced its offtake from the minimum given in its contract of 300 Mcf/d to 180 Mcf/d. A reason was excess supply in the United States, which succeeded in pressing down the price of gas purchased from Mexico from 4.99 dollars/MBtu to 4.40 dollars/MBtu, but not as far as specified in the original price-determining formula. This formula was:

$$\text{Gas price} = \frac{\text{Average price of 5 crudes}}{27.444} \times 3.625$$

In this expression, 27.444 is the average price of crude oil at the time of the agreement, and 3.625 the agreed price base. At one time Border Gas considered this formula unpalatable, particularly when it appeared that the price of oil might rise to extreme heights; but with the weakening of the oil price that began in the early 1980s, Mexico found it unnecessary to protest the fall in the demand for gas by the United States, particularly since there was also a growing demand for gas in Mexico. In addition, although on the surface it appears that within the non-OPEC developing countries Mexico has the largest export potential, it has been claimed that if the associated gas reserve in the undeveloped Chicontepec field was placed in the deferred reserve category, Mexico's exportable surplus would be quite modest.

The basic problem for Mexico is similar to that facing Norway and Algeria. After building up expectations for a sellers market, these countries were faced with the opposite. Instead of being able to sell their gas at about 4.95 dollars/MBtu, as they had planned, both Mexico and Canada have recently found themselves confronted with market prices for gas in the United States that are under 3 dollars/MBtu, and occasionally well under. This is because, for the past few years, there has been a gas bubble in the United States, and forecasts are that it will remain throughout most of this decade, if not longer. Canada has shown itself flexible and lowered the price of its gas, while Mexico — which has much more to lose in the long run if wrong decisions are made about their irreplaceable hydrocarbon resources — has decided to be cautious and sell less. To my way of thinking this is very understandable, because both gas and oil have proved to be a mixed blessing for Mexico. They have created wealth but brought inflation and corruption.

With the exception of Mexico, it is difficult to foresee a large exportable surplus of gas in Latin America — with the possible exception of Argentina. In the Caribbean, Trinidad may eventually be able to export large amounts of gas. The only other Latin American countries that may be able to generate an exportable surplus of gas are Bolivia and Colombia.

APPENDIX: PRICES AND DEREGULATION

It is not economic theory, but common sense which tells us that deregulating the gas price will probably cause it to rise. However in the deregulation of United States gas an interesting situation arose in which, as deregulation progressed, the price of gas fell.

Soon after the passage of the Natural Gas Policy Act (NGPA), 95 per cent of US natural gas was regulated and 5 per cent (so-called high cost gas) was exempt from price controls. Where the unregulated category was concerned, price could rise to clear the market. At the regulated price there was considerable excess demand.

Using this arrangement as a starting point, further deregulation resulted in a lower price. It was inevitable, because deregulation resulted in an increased supply (Figure 4.1).

Figure 4.1: Price changes as deregulation takes place

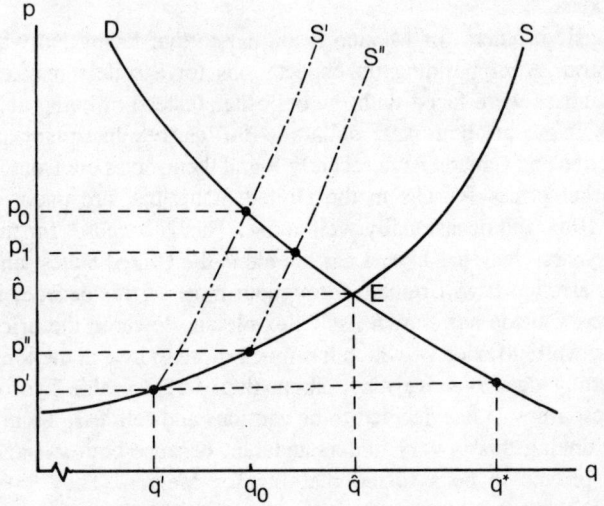

First, let us examine the situation described in the second paragraph in this appendix. The regulated price is p', and at this price the amount of gas available is q'. Excess demand at this price is $q^* - q'$. By introducing a supply of (and a supply curve S' for) high cost gas, the market clearing price becomes p_0. Observe that in this market there are two prices, p' and p_0. The average market price is:

$$p_a = \frac{p'q' + p_0(q_0 - q')}{q_0}$$

The five per cent mentioned above corresponds to the percentage represented by $(q_0 - q')/q_0$.

Next, we can examine what happens when the price is deregulated. For instance, we raise it from p' to p''. By displacing the supply curve for high cost gas to S'', the market clearing price drops to p_1. What about the average price? We have $p_1 < p_0$, but $p'' > p'$, and so it appears that the average price can rise *or* fall; but eventually it has to fall if deregulation continues. To see this we have only to suppose that we deregulated all the way to the price $\hat{p}$, which gives us $S = D$ (at point E). Then there is no question as to what has happened to the average price. It is $\hat{p}$, and $\hat{p} < p_0$.

5

Soviet Natural Gas and the Western European Energy Crisis

This chapter examines the Soviet natural gas sector and its relationship to the energy requirements of Western Europe. There is also an appendix on the present tribulations of the Soviet oil industry, but I start with background information about the Western European (and the world) economy. See also J.F. Stern (1981, 1984).

Since 1973 the economies of the Western industrial countries have been in serious difficulty much of the time; and nowhere has the agony been more profound than in Western Europe. If we overlook a lack of imagination on the part of some governments, as well as the application of inappropriate economic theories (such as monetarist economics) to real life situations, then one of the most important causes of this dilemma was the oil price shocks of 1973 and 1979–80. In 1973 the annual growth of non-residential fixed investment in four leading industrial countries (United States, Japan, Germany, and the United Kingdom) left its long run trend (of 4.9, 15.5, 6.5, and 5.8 per cent respectively) and fell to a lower trajectory, while inflation increased. The fall in productivity that was so pronounced in certain industrial countries (for example the United States) can also largely be attributed to the oil (and energy) price rises that have taken place since 1973, principally because of the reduction in investment and, on the psychological side, the lowering of expectations and morale caused by the adjustment to these price shocks. In the OECD, unemployment has increased by about 14 million persons in the past 11 years, and now stands at almost 30.5 million. Unemployment has been as high as 33 million and according to some analysts it will reach that figure again in the near future. Needless to say, Western Europe will find itself supporting a large part of this unemployment.

In order to restore even a semblance of the normality that existed prior to 1973, an adequate supply of energy must be available to the industrial world: with the exception of bringing wages and salaries into line with productivity, this is the most important economic issue facing the Western industrial democracies. There are many good sides to the decline in energy use that we have witnessed over the past five or six years, but things have happened that are definitely unfortunate. Among these are the fading away of a portion of the energy-intensive manufacturing sector of North America and Western Europe, and the loss of hundreds of thousands — if not millions — of high-paying industrial jobs. Some who worked in these industries have transferred to high technology, but most are now associated with low-tech industries and activities. Many of the high technology activities that were supposed to replace industries such as aluminium refining, steel making, and ship building will not materialise. This is because these high technology industries will be established in low-wage, non-unionised parts of the world which now, as was not the case in the fairly recent past, are in possession of a large and growing reservoir of technical and managerial skill.

THE GREAT GAS TRANSACTION – I

We can now turn to some of the economic and political aspects of the arrangement that some Russian gentleman has termed 'the deal of the century', around which an extraordinary number of misunderstandings took shape. The first has to do with the so called 'intolerable' level of Western European dependence on Russian natural gas, when in fact the axis of dependence runs in the opposite direction. The Soviet Union has evidently decided that in the future its economic development will be promoted through the increased import and application of foreign technology — and not, for example, through wider application of capitalist incentives of the type associated with the once mooted Liberman reforms. This technology can be acquired only if 'hard currency' is available to pay for it, which means that the Soviet Union has no choice but to export a great deal of oil and/or gas to the capitalist world. (I would not be surprised to see the Liberman reforms resurrected by the present Soviet government.)

For reasons given in the previous section, Western Europe must have access to large, reasonably priced and secure flows of energy if its deindustrialisation is to be checked; and as will be argued below, natural gas may be the ideal medium. Western Europe is not constrained to take this gas from the Soviet Union. A great deal of natural gas can be obtained from less controversial suppliers than the USSR. I am speaking of course of the North Sea gas nations where reserves in the Norwegian sector seems to be expanding all the time; and also pipeline gas from Algeria. Moreover, other types of energy resources are readily available, though in many respects they are not as satisfactory as Soviet gas. These other resources include coal from North America and Europe, nuclear power, and eventually a great deal of liquefied natural gas from suppliers outside of Europe.

Bargaining power lies in the West. It has already been used to obtain important concessions from the Soviets on both the price of gas and the interest rate on the money that the Soviets will borrow to finance the purchase of pipe, pipelaying machines, and equipment for compressor stations. The price of Soviet gas has proved to be considerably lower than the price of other gas being purchased in Europe. Another factor of importance to the present discussion is that a large part of the gas consumed in Europe is used in electricity plants and industrial boilers where it is a comparatively simple matter to switch between gas and oil: if for one reason or another consumers experienced serious problems with Soviet gas, they would experience few technical or economic problems in switching back to oil, although perhaps enough problems to cause the Soviets to lose an appreciable part of the Western European market forever. As pointed out below, there have been interruptions in Soviet deliveries of energy materials to Western purchasers in the past, and every intelligent person who observes the antics of politicians in East and West Europe must realise that rationality is in short supply at times in most European governments; but I remain convinced that the Soviet government will do everything possible to expand deliveries of natural gas to Western Europe, which means that they will do everything possible to avoid future interruptions.

It might also be useful for Western European gas purchasers, and potential purchasers, to remember that when gas is available in large quantities, and can be transported by pipeline to the centre of industrial and urban areas it may be the most efficient of all energy resources. This is because large flows permit important economies of scale in the operation of pipelines, compressors, etc; while large reserves ensure that the lifetime of this equipment is fully utilised. This has proved true in both the Soviet Union and the United States. In the USSR, gas consumption went from three per cent of total world consumption in 1950 to 24 per cent in 1978. During the same period the expansion in the Soviet pipeline system could only be called phenomenal, and a similar story took place earlier in the United States, where an even more efficient natural gas network was established. There is also reason to believe that further important discoveries of conventional natural gas will be made in the Soviet Union, especially when the vast areas of arctic Russia can be explored. In carrying out this gigantic task there will be a huge demand for Western European offshore expertise, and already there have been discussions between the Soviet energy authorities and offshore enterprises in Scandinavia. One would like to feel that this will turn out to be a glorious opportunity for at least a partial revival of the Swedish shipbuilding industry, which is already engaged in offshore activities, but unfortunately the eyes of the present Swedish government are permanently turned south, and to giving away money instead of thinking about the Swedish economy.

Next, there is a delicate political issue. Although it is painfully obvious to those of us interested in such things, a perverse gentleman's agreement requires this issue to be overlooked or played down on the more pedestrian academic turf: although they are political allies, the United States and most of the states of Western Europe are economic rivals. They are rivals now, always have been, and in my opinion will continue to maintain this relationship in the foreseeable future — though at an increased intensity, because they will not only be competing with each other, but with Japan and the Newly Industrialising Countries of East Asia. Moreover, as opposed to Western Europe, the United States is energy rich. This fact should have been brought home to many observers at the onset of the two oil price shocks, when the American dollar appreciated dramatically relative to most currencies as the foreign exchange markets payed their tribute to the political and economic significance of North America's energy potential.

Since Western European technology is at least on a level with that of the United States, access to Soviet energy supplies could make the difference between prosperity for Western Europe or a gradual slide into pauperdom. Consider the European steel industry, which is in the process of being truncated by the so called D'Avignon plan. Soviet orders for steel pipelines and ancillary equipment did not bring the bloom of springtime back into the cheeks of the ailing Western European industry, but these orders were important, and future orders could be even more important. There is also the possibility that an outright rejection of Soviet gas on political grounds would mean that trade that might be carried out with Western Europe would then take place between the Soviets and the countries of East Asia.

How would more intensive contacts between Western Europe and the Soviet Union affect the other states of Eastern Europe and, for that matter, the United States? Several of these countries (e.g. Poland and Rumania) have been operating on the rim of bankrupcy, since under the joint economic plan for Eastern Europe they tended to pursue an excessive degree of specialisation. More economic contact between Eastern and Western Europe would help promote the rationalisation of the production structures of the Eastern European countries, since a wider range of trading opportunities with the West would help break down the present inefficient or non-optimal pattern of inter-country specialisation practised in the East. These trading opportunities could also mean such things as cheaper food and expanded opportunities for tourism for Western Europe. *Ceteris paribus* the United States should also gain if the European economy were to regain its momentum. It is for this reason that many multinational firms, particularly among the major oil companies, have supported the Soviet-Western European gas transaction even though it did not involve any immediate benefits for these companies or their affiliates.

Finally, a few words about the initial opposition of President Reagan to expanding the level of gas imports from the Soviet Union into Western Europe. (Note that I say President Reagan, and not the Reagan government, because many — if not most — members of the President's cabinet were against any palpable economic opposition to 'the deal of the century'. This is why, after a decent interval, his opposition faded.) What President Reagan and some of his advisors did not understand is that the basic issue is not politics but economics: safeguarding and expanding Western Europe's energy supplies in the years to come, and widening markets in the East. Some European politicians and political scientists understand only too well that the rearmament taking place in the United States is at least partially intended to provide a screen or deterrent to prevent Russian interference with United States military operations in the Third World. For reasons that are only too obvious, some of these operations could involve the Middle East, and might lead to another OAPEC boycott of Western European oil of the 1973 variety. In this type of situation, the more energy that is available to Western Europe the better.

THE GREAT GAS TRANSACTION – II

Energy trade between Western Europe and Russia dates from the last century, although after the Second World War Soviet energy exports were insignificant until about 1965. Since that time substantial contracts have been signed for the export of Soviet oil and gas, despite occasional political turbulence between the Soviet Union and the Western allies. For instance, in 1982 — at a time when Prime Minister Mitterand was eloquently reaffirming the loyalty of France to Nato and its ideals — France signed a new 20–25 year contract for 8 billion cubic meters per year (= 8 Gcm/y) of Soviet gas, with the maximum flow to be reached in 1986–7, although contract provisions permit France to reduce deliveries by 20 per cent with no penalty. Forecasts suggest that France's dependency on the Soviet Union for energy will be about 4.5 per cent in 1990, and gas dependency will be about 35 per cent. Corresponding figures for Italy are 5.6 and 30 per cent, and for West Germany 5.3 and 30 per cent — although I have seen predictions which indicate that the West German figures are too low and should be 6 and 40 per cent.

Originally the Soviets were scheduled to provide Western Europe with an additional 40 billion cubic meters per year (= 1.4 trillion cubic feet per year, or 1.4 Tcf/y) of natural gas for at least 20 years. Deliveries in 1982 came to about 27 Gcm, and it was estimated that Soviet revenues from these sales came to about 3 billion dollars, which indicates a price of 3.15 dollars/thousand cubic feet (or approximately 3.15 dollars/million Btu). The base cost of the gas in the new contracts is 4.60 dollars/MBtu, at the Czechoslovak frontier, and the approximate transport cost from that point to Paris will be about 0.30 cents. On the other hand, West Germany was supposed to pay an average base price of 4.75 dollars/MBtu for delivered gas.

The new pipeline originates in Tyumen Province, just north of the Arctic Circle, and one part of this project is an extension of the already constructed Ukhta-Uzhgorod, or Northern Lights pipeline, which crosses the Russian border into Czechoslovakia and proceeds to Austria and West Germany. The line is 56 inches (= 1.42 meters) in diameter, and the line pressure is 75 bar. Of the originally projected 40 Gcm/y, West Germany was to take 10.5 Gcm/y, France and Italy 8 Gcm/y each, The Netherlands and Belgium about 5 Gcm/y, Austria 3 Gcm/y, and Switzerland 1 Gcm/y. As things now stand these Western European countries have taken only about

50 per cent of the amounts they were originally scheduled to take because their demand for all energy inputs has declined. In addition, because the price of the gas they are buying is indexed to the price of various crude oils, they are now paying about 3.65 dollars/MBtu for this gas, which is extremely reasonable.

Let us get some idea of what the above price means in terms of the price of oil. Thermally, or in terms of heating values, 1,000 cubic feet of natural gas (= 1 MBtu) is the equivalent of 0.178 barrels of crude oil. Thus it would take $1,000/0.178 = 5,620$ cubic feet (= 5,620,000 Btu) to get the same energy content as one barrel of oil. At a price of 4.75 dollars/MBtu, this means that Soviet natural gas would be selling at an oil-equivalent price of about 27 dollars per barrel; at a price of 3.65 dollars/MBtu, the equivalent oil price would be 20.75 dollars/barrel (the present oil equivalent coal price is about 14 dollars/barrel). The price of Soviet gas will be allowed to fluctuate in phase with changes in the price of oil, and the index used will apparently be one in which the weights are as follows: light fuel (heating) oil, 40 per cent; heavy fuel oil, 40 per cent; and various crude oils, 20 per cent. There is also supposed to be a floor price. At the same time France was negotiating with the Soviet Union for gas, an agreement was reached by that country to pay Algeria an FOB price of 5.1 dollars/MBtu for the supply of 9.1 Gcm/y of LNG. The cost of transport raised the price of these gas supplies to almost 6 dollars/MBtu. Similarly, the cost of North Sea gas from the Statfjord field, delivered to Emden (West Germany) was to be 5.5 dollars/MBtu.

The indexing of the price of Soviet gas should be of interest to students of price theory, because the larger the amount of gas purchased, the greater the downward pressure on the price of oil which, through the indexing formulae, also places a downward pressure on the price of gas. For instance, when Soviet gas from the new pipeline began flowing into Western Europe in large amounts in 1984, it caused the price of gas to fall all over the world. It also caused a great deal of annoyance with some of us who had supported the use of more Soviet gas. The annoyance was on the part of a Scandinavian gas producer, and not gas consumers who prefer purchasing gas at less than 4 dollars/MBtu, rather than the 7 dollars/MBtu this particular producer was interested in obtaining. Something else that should be remembered is that increased revenues from gas could lead to greater availability of Soviet energy supplies in the future, since the decision makers in that country might find themselves with the financial resources to expand their investments in gas, oil and coal. A major expansion of the Soviet energy sector could result in large imports of steel and machinery from OECD countries.

There were other economic gains to Western Europe because of the purchase of Soviet gas. At one time the Soviets would have borrowed a large part of the money to build this pipeline at an interest rate of 9.75 per cent (= 7.75 per cent in interest charges plus a two per cent surcharge on the price of equipment delivered), but as things turned out they delayed making this arrangement, and had to pay more than 11 per cent on a ten-year loan. These loans are repaid from gas revenues, and thus can be regarded as high quality. This is comforting for the financial institutions that made these loans, many of which include in their portfolios billions of dollars in IOUs that will never be honoured. Furthermore, it is fairly widely appreciated in world banking circles that an increased Soviet cash flow may decrease the possibility of a major Eastern European debt default. This is because the Soviet Union has sold gold to assist the Polish government in meeting interest payments on its large external debt. Finally, it can be pointed out to economists and journalists who have insisted that the loans made to the Soviet Union were bad business, that since most of these loans were taken out in European currencies, and particularly Deutschmarks, the current decline in the value of the American dollar raised the interest rate on these loans measured in terms of the dollar to about 15 per cent. This puts them well above the European prime rate.

Some aspects of Western European energy dependence can now be discussed, beginning with Table 5.1.

Although less gas is currently purchased from the Soviet Union than envisaged in the new contracts, it is almost certain that at least the 40 Gcm/y originally discussed will be purchased, and probably a great deal more. For instance, I expect the amount of gas that will be purchased from the Soviet Union around the turn of the century to exceed the quantities shown in Table 5.1. Furthermore, I personally dismiss the prospect of the Soviets attempting to turn these gas exports into political leverage by either the threat of a boycott or an actual boycott. There is too much gas outside the Soviet Union for the Russians to have any sort of monopoly, because the Western European pipeline grid assures that gas from any source can be moved to almost every important urban and industrial centre in Western Europe; and most important because the Russians have no intention of turning their back on the large amount of cash that can be earned by exporting gas to Western Europe, and this is particularly

Table 5.1: Forecast supply and demand of natural gas in the European Community for 1990 and 2000 (in million tonnes of oil equivalent — mtoe)

	Total gas consumption		Total gas production		Imports from Algeria		Imports from Norway		Imports from Soviet Union	
	1990	2000	1990	2000	1990	2000	1990	2000	1990	2000
West Germany	60.0	57.5	14.0	>9	–	–	7.8	2.8	15.8	17.3
France	26.3	23.0	2.3	1.3	7.7	4.9	3.4	1.7	9.5	9.5
Italy	32.9	37.0	6.3	>4	8.6	10.9	–	–	10.0	11.0
Netherlands	26.6	27.6	52.0	40.3	–	–	1.6	0.9	–	–
Belgium	8.5	10.5	–	–	3.7	4.2	2.3	0.8	–	–
Luxembourg	0.6	0.7	–	–	–	–	–	–	–	–
UK	45.4	48.0	33.7	30–40	–	–	11.7	<11	–	–
Ireland	1.5	1.8	1.7	2.0	–	–	–	–	–	–
Denmark	1.6	2.0	2.0	2.3	–	–	–	–	–	–
Greece	0.1	–	0.1	–	–	–	–	–	–	–

Note: These figures were obtained by averaging various estimates and should be considered tentative. For example, the Soviet Union could be selling gas to the UK by the year 2000 or, on the other hand, the figure for imports from Norway could be doubled that given here because of the recently concluded arrangements concerning the Troll field.

true at the present time when they are uncertain about the future of their oil production. They also know that if they build up an appreciable Western European dependence on the Soviet Union, and then reduce gas deliveries for political reasons, it is almost certain that every possible measure will be taken to cut back on the use of Soviet gas, and these measures will be successful — though costly.

There have been some reductions in the delivery of energy supplies from the Soviet Union in the past, and on some occasions these appear to be politically motivated; but none in the recent past. In 1980 the Soviets reduced deliveries of anthracite coal to France in order to compensate for a drop in output at Polish mines; and during the winter of 1980–1 there was a cutback in contractual deliveries of gas to Western Europe that, the Soviets said, was related to the cold weather in Siberia. In the first of these cases it can be said that no one, anywhere, has shown that the reduction in Soviet deliveries was intended as a political move against France; while in the second, much of this gas was delivered later. On the other hand there have been instances when the interruption of Soviet energy supplies was clearly politically motivated: the curtailing of oil deliveries to Yugoslavia in 1958, Israel in 1956, Finland in 1958 and China in 1960. However Yugoslavia is currently an important purchaser of Soviet gas (Table 5.2) while Finland in 100 per cent

Table 5.2: Soviet natural gas exports to Western and Eastern Europe (Gcm)

Country	1973	1975	1980	1981	1982	1984	Dependence[a]
Western Europe	2.0	8.0	26.0	29.2	26.7	31.7	
West Germany	0.4	3.1	10.8	11.8	10.5	13.6	(25.4%)
France	0.0	0.0	4.0	4.6	3.8	4.9	(17.8%)
Italy	0.0	2.3	7.0	8.0	8.8	8.4	(28.4%)
Finland	0.0	0.7	1.0	0.8	0.7	0.7	(100.0%)
Austria	1.6	1.9	3.2	4.0	2.9	4.1	(92.9%)
Eastern Europe	4.1	11.2	29.4	29.9	31.1	31.9	
Bulgaria	0.0	1.2	4.0	4.5	4.8	4.8	
Czechoslovakia	2.4	3.7	8.3	8.6	9.0	9.3	
Germany Democratic Republic	0.0	3.2	6.4	6.3	6.4	6.4	
Poland	1.7	2.5	5.3	5.3	5.6	6.0	
Rumania	0.0	0.0	1.5	1.5	1.5	1.5	
Hungary	0.0	0.6	3.8	3.8	3.7	3.9	
Yugoslavia	0.0	0.0	0.0	2.0	2.4	3.0	
The world	6.8	19.3	57.1	61.2	60.3	66.7	

[a] Preliminary figures
Source: Economic Commission for Europe and Deutsches Institut für Wirtschaftsforschung, (annual reports (1984, 1985)

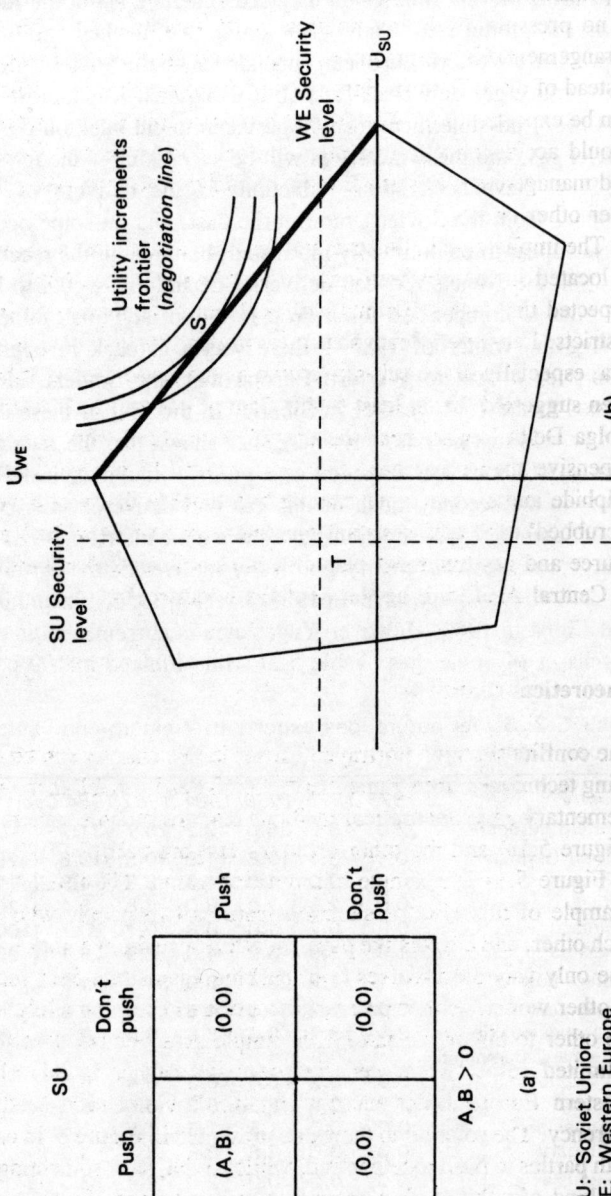

Figure 5.1a and b: Coordination and Nash-type games

Note: SU: Soviet Union
WE: Western Europe

dependent on the Soviet Union for gas, and as far as I know there is no pressure from any political party in Finland to change this arrangement. However, given attempts by the Soviets to export gas instead of oil to Eastern Europe (so that the much more valuable oil can be exported to the West), if it happened that Soviet gas supplies should accidentally fall because of such factors as bad weather or bad management, Eastern Europe would probably be given priority over other clients.

The implication in this chapter has been that almost all Soviet gas is located in the northern part of the Soviet Union; however it is expected that large amounts will eventually be discovered in other districts. Large petroleum structures are known to exist in the Black Sea, especially in the sub-sea known as the Sea of Azov. It has also been suggested that at least 6,000 Gcm of gas will be located in the Volga Delta Region near the city of Astrakhan. This gas is more expensive than other gas, since it contains 25 per cent hydrogen sulphide and 10 to 12 per cent carbon dioxide that will have to be 'scrubbed', and acid-resistant pipeline may be needed between gas source and gas treatment plants. It is also thought that Turkmenia, in Central Asia, will be the next gas boom region.

Theoretical comment

The conflict situation portrayed earlier in this chapter can be treated using techniques from game theory. As far as I can tell there are two elementary game theoretical methods that are suitable; one is trivial (Figure 5.1a) and the other arbitrary (Figure 5.1b).

Figure 5.1a is a simple coordination game. The usual textbook example of this kind of situation features two people who dislike each other, and also dislike physical effort, caught in a house on fire. The only way out involves both pushing against the door together. In other words, neither possess the option of escaping while leaving the other to his or her fate. This simple arrangement describes, in a limited respect, the scenario presented earlier in this chapter: Western Europe needs energy and the Soviet Union needs hard currency. The solution to the game presented in Figure 5.1a calls for both parties to push together and, while trivial, says something about the kind of solution that appears in many games characterised by actual or latent conflict, and which are in no way trivial: it usually pays to cooperate in situations where cooperation is possible. To this can be added, when cooperation is not possible, then steps should be taken to make it possible.

The other approach is the Nash arbitration scheme. Here, no deal between SU and WE results in a non-optimal solution that is shown by T. This is sometimes called the threat point, and it shows the damage that each side can inflict on the other by *not* reaching an agreement. The optimal solution is at S. Any solution involving the Nash preference system is arbitrary, because given the well-known impossibility of aggregating preferences, there is no reason to believe that Nash's preference scheme (which is shown in the usual form by indifference curves) is superior to a number of others.

Given that the possibility set is convex, and the Nash indifference curves are quasi-concave, we have the usual type of neo-classical optimisation exercise which results in the usual solution. Perhaps the most interesting aspect of this arrangement is the importance of the 'security', 'threat', or 'status quo' point T. As argued in this chapter, the bargaining advantage is actually with Western Europe, and so in the diagram the security point shows a positive utility level for WE, which as shown has an important significance for the extent of the payoff frontier that is relevant. The payoff frontier is the straight line stretching between the U_{WE} and U_{SU} axes, while the part that is relevant is the negotiation line (which is the thicker part).

ALTERNATIVES TO SOVIET GAS

We can now consider several alternatives to Soviet natural gas. The pipeline from Algeria, through Tunisia, to Italy — whose terminus is Bologna — has already been mentioned, and already there are plans to greatly extend its capacity and to prolong it into Central Europe. A pipeline between Algeria and France with a terminus in the vicinity of Marseilles has also been discussed for many years, and also a line from West Africa (Nigeria and the Cameroons) to Europe, as well as LNG projects involving these two West African countries. As previously noted France receives LNG from Algeria, and the price of this LNG is indexed to a basket of light fuel oils.

The interesting thing about the last transaction is not just the base price that will be paid for this gas, which is appreciably higher than the base price of Soviet gas, but the fact that Algeria has distinguished itself over the past few years as the champion of a *prix juste* — that is, a gas price that is at least as high as the oil price; and they have been strongly supported by Norway. This attitude has caused bad feeling between Algeria and its customers and potential customers. There is also a basic economic reality about all these existing and potential LNG deals that many economists insist upon ignoring: as long as pipeline gas from the Soviet Union is available under the present conditions, the purchase of LNG from anywhere must involve the payment of a subsidy.

For obvious reasons, oil need not be taken up here, and the matter of nuclear energy can also be treated rather quickly. From an economic point of view I am convinced that, in both theory and fact, only hydro-electricity can compete with uranium or thorium as a source of inexpensive electricity, but the political and environmental costs may be prohibitive. Nobody, anywhere, has proved that there is a satisfactory solution to the problem of disposing of nuclear wastes; and it also may be true that the successful operation of conventional nuclear equipment will help smooth the way for the widespread operation of breeder reactors. Clearly, this last alternative is unacceptable on a number of grounds, the principal of which is that if a mistake is made, it might turn out to be a mistake that — at the present level of technology — cannot be rectified. Furthermore, the security problems that are associated with a full fledged 'plutonium community' may be such that they are not manageable under the conventional forms of democracy that are found in Western Europe.

That brings us to coal. Unless I am mistaken, I was one of the first economists in the world to suggest publically — after the first oil price shock — that coal should be returned to its former position as the industrial world's most important energy resource. As a long run option this may make sense, but in the short run great care will have to be taken: the results of raising the consumption of coal by a factor of two or three in the next 20 years could be catastrophic from an environmental point of view. This is so because in the present climate of recession, wavering profits, and increased defence and welfare expenditures, greater efforts are being made everywhere to avoid making investments of the type needed to turn coal into an environmentally acceptable commodity, and many of these efforts are successful.

I would like to conclude this section by mentioning two reasons that were occasionally mooted for the initial opposition of the Reagan government to the introduction of larger quantities of Soviet natural gas into Western Europe. The first was the unvarnished desire to deprive the Soviets of the foreign exchange they needed to import foreign technology and equipment, and thereby raise their rate of economic growth.

It has also been suggested that, for commercial reasons, many people in the United States would like to see Soviet gas replaced by United States coal. In many respects, this would not be good business for Europe, and in the long-run Soviet coal might even prove to be less expensive than United States coal. However, given the political and social dynamite present in the unemployment rate now prevailing in the OECD, a large increase in the production and export of United States coal might also turn out to be good for Western Europe, particularly if it helped prolong the present economic upswing in the United States.

CONCLUSION

This chapter has reviewed economic and, to a limited extent, political problems associated with the supply of Soviet natural gas to Western Europe. My conclusion is that the more of this gas purchased the better. What I have not pointed out is that once the tap is fully opened on Soviet gas — and it shows signs of staying open — it may tie the huge gas reserves of Iran into the European gas consumption picture via pipelines through Southern Russia.

As for the politics of buying Soviet gas, I have only this to say. Having worn the uniform of the United States Army in Western Europe in the early 1950s, the only kind of attack on Western properties by Soviet citizens at the present time that I can imagine would be an attack on the duty free gin and whisky so prominently displayed at Western diplomatic and commercial receptions. Also, since the concept of Western European energy dependence on the Soviet Union is at best a flimsy myth, it is in the interest of Western European households and industries that negotiations with the Soviets begin as soon as possible for the huge amount of natural gas that Western Europe will require later in the century, and to do so before some bright politician gets the idea that it would be better to pay a price that is 25 to 50 per cent higher for gas from some other source. There are plenty of these ideas circulating in the corridors and restaurants of power, particularly after the cognac has gone around the table once or twice.

The simple fact is that Western Europe has everything to gain and nothing to lose — either politically or economically — from buying as much gas as possible from the Soviet Union.

APPENDIX: SOVIET OIL INDUSTRY

In 1984 Russia produced 613 million tonnes (= 12.3 Mbbl/d) of oil, and exported an estimated 90 Mt of crude oil and oil products to the World Outside the Centrally Planned Areas (WOCA). The exact pattern of Soviet oil exports is shown in Table 5.3. These sales brought in about 19 billion dollars of total Soviet exports of 30 billion dollars. In 1984 the Soviet Union supplied Western Europe with almost 15 per cent of its total requirements of crude oil, and supplanted Saudi Arabia as the largest supplier of oil to the European Community.

Table 5.3: Soviet exports of oil (millions of barrels per day — Mbbl/d)

Destination	1973	1975	1980	1982	1984
World	2.73	2.61	3.26	3.27	3.53
Western Europe	0.93	0.96	1.21	1.33	1.57
West Germany	0.13	0.17	0.14	0.17	0.16
France	0.11	0.07	0.17	0.14	0.16
Britain	0.02	0.03	0.03	0.06	0.07
Italy	0.17	0.14	0.17	0.17	0.24
Netherlands	0.06	0.06	0.15	0.25	0.27
Finland	0.20	0.18	0.19	0.21	0.21
Other Western Europe	0.33	0.36	0.45	0.38	0.39
Japan	0.04	0.03	0.01	0.01	0.02
United States	0.00	0.01	0.00	0.00	0.01
Centrally Planned Economies	1.35	1.55	1.92	1.78	1.76
Bulgaria	0.19	0.23	0.28	0.25	0.25
Czechoslovakia	0.29	0.32	0.39	0.33	0.33
German Democratic Republic	0.26	0.30	0.38	0.35	0.34
Poland	0.25	0.27	0.32	0.30	0.30
Rumania	0.00	0.00	0.03	0.01	0.03
Hungary	0.13	0.15	0.18	0.16	0.15
Cuba	0.15	0.16	0.19	0.19	0.18
Mongolia	0.01	0.01	0.02	0.02	0.02
Vietnam	0.00	0.01	0.01	0.04	0.04
Other	0.09	0.11	0.12	0.12	0.12
Less Developed Countries	0.06	0.09	0.12	0.16	0.19
Afghanistan	0.00	0.00	0.01	0.01	0.01
Ethiopia	0.00	0.00	0.01	0.02	0.02
Brazil	0.00	0.03	0.00	0.02	0.01
India	0.01	0.02	0.08	0.10	0.12
Other	0.05	0.03	0.01	0.02	0.03

Note: Figures include oil and oil products.
Sources: *The Petroleum Economist* (various issues); Deutsches Institut für Wirtschaftsforschung.

When Soviet oil production fell by four per cent in the first five months of 1985 — and this was the first time that the output of crude oil had declined since the Second World War — it was judged a serious matter, especially when the fall in oil prices reduced the revenue obtained from every barrel of oil: at the beginning of June 1985, the official price of Urals crude was reduced from 28 dollars/bbl to 27 dollars/bbl, and spot sales were made at 26 dollars/bbl. It also become apparent, about the middle of 1985, that the 1985 production target of 628 million tonnes could not be reached, regardless of the number of managers in the oil sector that were relieved of their duties.

Still, the Soviet Union is in no danger of an energy shortage, and there is little reason to believe that Soviet oil revenues will collapse unless the price of oil collapses. In 1984 export revenue was maintained because, like the countries of the capitalist world, the Soviet Union held large inventories of oil that could be reduced; in addition oil received from Middle Eastern countries in return for Soviet arms could be re-exported. The Soviet Union has the same problem as the other oil-exporting countries, in that it cannot export too much oil without materially contributing to downwards pressure on the price. On the other hand, if production in the Soviet agricultural sector can be increased over the disappointing levels of the past few years, less hard currency will be necessary; present predictions are that there will be a substantial increase in grain production in the years to come.

The good news on the Soviet energy scene (for the Soviets at least) is in the gas sector. Soviet gas production increased ten per cent to 587 Gcm in 1984, which was well above the figure given in the latest five year plan; and output was 473 Gcm in the first nine months of 1985. There is also extensive substitution of gas-for-oil taking place in Soviet power stations, and since about 100 Mt/y of fuel oil is used in this activity, this may have saved as much as 10 Mt in 1984. Expectations are that another 10 Mt/y will be saved by 1990.

As I understand the situation, much more effort is to be put into raising the output of crude oil, and in particular there is to be a major increase in deep drilling, particularly in Western Siberia. It is difficult to say exactly what this is going to mean for the Soviet oil industry, but from a geological point of view it cannot be judged as favourable to the Soviet Union. As in the United States, the emphasis has shifted from giant or super-giant oil fields to deposits containing only a relatively modest amount of oil: for example, production peaked at 155 Mt in the vast Samotlar oilfield in 1980. Also, as is the case in the United States, Soviet oil reserves are not expanding in step with the production of oil; but even more serious, some of the older Soviet fields may be in a serious condition because of factors such as water flooding, low reservoir pressures, cracked oil pipelines (which sometimes cause fires and explosions) and badly corroded pipes. All in all, the cost of finding and extracting more oil is ascending rapidly, and the question is how much of the limited investment funds that are available for energy industries should be allocated to oil. The oil sector already obtains 19 per cent of the capital investment going into Soviet industry as a whole, and to give it more would mean throwing the rest of the Soviet plan out of balance.

Mikhail Gorbachev has decided to pay close attention to the oil industry because, among other reasons, he can no longer ignore the widespread opinion in the country that the basic problem in the oil industry is its top management. Some of the former stars of the Soviet oil industry have already been told to devote more time to their summer houses and vodka, and leave the running of the industry to managers that are capable of and interested in reducing the tremendous waste that plagues most Soviet industrial operations. The root of the trouble in the Soviet oil sector is the over rapid exploitation of the West Siberian oil reserves in the 1970s, and this cannot be corrected overnight, regardless of the managerial talent that can be placed at the disposal of the oil industry. The intention at one time was to raise the output of Western Siberia alone to 10.5 Mbbl/d, but instead output remains at about 7.6 Mbbl/d (out of a total output of 12.3 Mbbl/d), and may be slightly less at the time of writing. It is also said that one-fifth of the wells in Western Siberia are not working because essential repairs are needed.

Just as there is a large programme underway in the Soviet Union to substitute gas for oil, no stone is left unturned in the rush to use more nuclear power. Operating capacity in the nuclear branch of the Soviet electricity generating sector has now reached 21,000 MW, or 7.5 per cent of the total electricity generating capacity. During the twelfth Five-Year Plan (1986–1990), nuclear power is scheduled to increase at a rate of 10,000 MW/y, with the target for 1990 of 65,000–70,000 MW of capacity. The Soviets claim that each 1,000 MW of nuclear capacity in operation saves the equivalent of two million tonnes (= 2 Mt) of oil or 3 Mt of coal a year, and that large scale expansion of nuclear power will probably have to continue until the year 2050. There have been some problems at the

Atommash nuclear engineering works, which is supposed to become the principal equipment supplier to the nuclear industry, but it is claimed that in 1986 Atommash began turning out seven 1,000 MW reactors per year. Stations being built at Khmelnitskoye and in the northern part of the Ukraine will also supply electricity to Soviet allies in Eastern Europe, and Eastern European technicians are involved in their construction. Similarly, the output of coal is scheduled to rise, although the expected output in 1985 of 726 Mt is far short of the 770–800 Mt goal for 1985 that was given in the eleventh Five-Year Plan. Coal now provides about 28 per cent of domestic energy, and is scheduled to rise; but some of the most promising coal basins in the Soviet Union are still underdeveloped or undeveloped, and will remain that way until a great deal more money is invested in them — money which is unavailable at the present time. New developments in Soviet coal include increased efforts to install large capacity coal slurry pipelines, and the attempt to raise production and exports in the Far East with the help of Japanese loans. Fairly large coal gasification and liquefaction plants are also in the advanced planning stage.

Despite occasional setbacks, the Soviet Union is bracing itself to dominate the Western European energy market by the end of the century. Demand for the gas from the Urengoj-West European export line has been disappointing, with the originally projected capacity of 40 Gcm/y downgraded to 32 Gcm/y, and only 26 Gcm/y sold; but by the end of this decade that route may be carrying as much as 40 Gcm/y, particularly if the performance of the Western European economy does not worsen. Just now there is an upswing in the Soviet economy, with a macroeconomic growth rate that may be higher than that being experienced in Western Europe, and productivity per worker increasing by about three per cent/year. It may be that the Soviet economy has entered a new phase, which would surprise some observers; although I would be surprised if the best endowed country in the world in minerals and energy could not provide most of its citizens with a standard of living at least as high as industrial countries that have no such resources.

Finally, the recent accident at Chernobyl needs to be commented on. The main effect of this accident on Soviet energy planning is its cost: altogether, between five and ten billion dollars may have been lost that could have been used for other investments in the energy sector. It has been specifically stated at the highest political levels that the expansion of nuclear energy will continue as planned. What will not be allowed to continue are mistakes of the kind that led to this accident — some of which might be classified as pure blunders, and some of which are analogous to 'pilot error'. My own belief is that countries with a high level of technological efficiency that are geographically proximate to the Soviet Union (such as Sweden) should accelerate their research on nuclear safety, and then take every step possible to transfer this technology to the Soviets and to others who might need it, even if it must be transferred free of charge.

6

OPEC and Other Developing Countries

BACKGROUND

Many OPEC countries are now aware that their supplies of hydrocarbons are not inexhaustible. They sense the contradiction of short run developments by long run trends, which means an oversupply of gas (or oil) now, as compared to possible shortages in the medium to long term. But even so studies are going ahead concerning the feasibility of export projects that could represent drastic changes in the world natural gas picture: pipelines from West Africa to Europe; huge gas deliveries in the form of LNG from Iran to Japan and Western Europe; the development of the unassociated gas in the giant North West Dome gasfield in Qatar; and so on.

While OPEC possesses 35 per cent of the world's gas reserves (with 25 per cent of the world's gas reserves in the Middle East), many OPEC countries have only limited supplies of gas, and this latter category includes some Middle Eastern countries. The countries that are capable of exporting large quantities of gas at the present time are Algeria, Indonesia, Abu Dhabi (of the UAE) and perhaps Libya. Countries that should join these are Qatar, Saudi Arabia, Nigeria and Iran. Saudi Arabia and Iran, especially the latter, have a long term capability as gas exporters; and when the large scale development of Iran's natural gas reserves begins, the world gas supply picture may be totally changed.

There is also considerable domestic consumption of gas within OPEC. In the Gulf there has been a sustained build-up of new petrochemical and fertiliser plants, power stations and desalinisation units that use gas as their source of energy. There is also substantial production of natural gas liquids, particularly in Saudi Arabia, where capacity may have reached 700,000 bbl/day. In many respects Saudi Arabia has been the most successful gas developer in the Middle East, and its achievements since 1973 have been in the world class. In 1975 Saudi Arabia initiated the first phase of a 14 to 15 billion dollar Master Gas Gathering System that is scheduled to provide 140 Mcm/d of gas when operating at full capacity. This system is for associated gas. In order to make the supply of gas independent of oil production, another system is being constructed to produce a larger amount of non-associated gas located in the Permian Khuff formation beneath the Ghawar oil field.

The Knuff formation is of interest to a number of countries in the Middle East, since it runs the entire length of the Gulf, from the UAE to Kuwait. Bahrain, for instance, is depleting its reserves of Khuff non-associated gas at about two per cent per year; 40 per cent of this is used for reinjection in Bahrain's oil field — that is rapidly being depleted; while the rest goes for power generation, running desalinisation plants, topping up the 'tail gas' used by Bahrain's aluminium smelter, and — after processing — as an input in Bahrain's refinery in the form of LPG. The LPG is produced in a small gas liquification plant that is 75 per cent owned by the government of Bahrain, and 12.5 per cent each by the Arab Petroleum Industries Corporation and Caltex. Caltex also markets 5500 bbl/d of LPG. The plant was constructed by the Japan Gasoline Co in 18 months for 100 million dollars, and this cost recouped in less than a year, although the original feasibility study indicated that it would take three to five years.

With the exception of Iran, Qatar and Saudi Arabia, the other countries of the Middle East — both OPEC and otherwise — have not been able to locate the gas they once thought was present. Kuwait, for example, despite its huge oil deposits, has actually experienced a gas shortage, with LPG and chemical fertiliser plants operating below capacity.

A country that does not have a problem with scarcity of reserves just now is Algeria, where proved energy reserves are at least 4 Gtoe, with gas comprising at least two-thirds. Probable and possible additional reserves amount to another 1.5 Gtoe. The basic problem for Algeria is to find large markets for its gas, because the European market on which it placed such reliance has been partially captured by the Soviet Union. In fact it was the appearance of large quantities of Soviet gas in Central Europe, with the possibility of much of this gas being routed to Italy, that changed Algerian opinions as to what comprised a suitable gas price. Algeria supplies pipeline gas to Italy via the TransMed pipeline, and also LNG to France, Belgium, Spain and the United States; but while Italy and the United States might eventually be interested in more gas, it is hardly likely that France and Belgium are interested in paying anything even approximating the prices that Algeria feels are fair. There has also been talk about an underwater pipeline to Spain of the type running to Italy. This may well come about but it seems unlikely that Algeria could win the price concessions from Spain on new contracts that they have managed to obtain from France and Italy.

It is also well to remember that Algeria is a country that has serious economic problems, despite the rosy outlook of a decade ago when it was discovered to possess large reserves of both oil and gas. Both reserves and exports of oil are slowly but surely falling, and while gas and condensates may be able to replace oil in the quantitative (i.e. volumetric) sense, it seems clear that gas is not going to bring in the kind of revenues that accrued to oil, especially since there is an oversupply of gas in the world. Furthermore, in the very long run, Algerian gas is exhaustible. If that country is not careful it could find itself in the unenviable situation of not having any hydrocarbons, nor much of anything else, because industrial development has not gone according to plan. Of course, Algeria is a huge country, and there is always the chance that there are still reserves of oil and/or gas to be found; but in my opinion this good fortune should not be reckoned with in the near future. More attention should be paid to the systematic and rational development of industry and agriculture, and reducing the high rate of population growth.

NATURAL GAS AND OPEC INDUSTRIALISATION

In Bahrain, gas is explicitly intended to be the basis of industrialisation. In addition to the Bahraini projects mentioned above, an iron ore pelletisation plant and a methanol plant have been initiated. The methanol plant is a link in the chain of projects planned by the Gulf Cooperation Council, in which the two key members are Saudi Arabia and Kuwait, and the products of this plant (methanol and ammonia) are to be marketed by SABIC (Saudi Arabian Basic Industries Corporation) and PIC (Kuwait's Petrochemical Industries Co). An important feature of this project is that it will be wholly Arab managed, and local technicians and managers will provide a complete line of engineering and administrative inputs.

Although many development economists have chosen to overlook the Middle East in their research, this part of the world is now experiencing the kind of textbook development that many of us, mistakenly it seems, expected all the resource-rich developing countries to undergo.

Just as Bahrain plans to use gas as an input for various industrial activities, Qatar has decided to sell most of its gas in primary form. One of the reasons for this is that Qatar has the largest single natural gas field in the world, and that country is also fourth in the world in reserves. Expectations are that when the North Field Project reaches full production capacity in the mid 1990s, Qatar will be one of the leading producers and exporters of LNG, although there has been some talk of exporting gas to Western Europe by pipeline. Naturally, projects of this nature are in the multi-billion dollar class, and it has been said that one of the reasons that Qatar occasionally exceeds its production quota for crude oil was to help finance some of the expensive investments it is in the process of making.

Natural gas exploration, production and marketing are currently supervised by the Qatar General Petroleum Corporation, but a new joint venture subsidiary has been formed to manage the export phase of the North Project. This subsidiary, Qatar Liquefied Natural Gas Co (QALIGAS) has British Petroleum and Compagnie Française des Pétroles as part owners: each has shares amounting to 7.5 per cent. Qatar General has the rest, but they have offered 15 per cent of their equity to a Japanese consortium if it will make a commitment to import half of the planned output of LNG, or about 3 million tonnes per year.

Something that has not been touched on here is the residential use of gas. As yet this use has not been fully developed in most OPEC countries, for several reasons. The first is that heating requirements in many OPEC countries are minimal as compared to the situation in the Northern Hemisphere, and there are already comparatively efficient cooking arrangements. Then, too, it makes more sense from the point of view of economic development to route as much gas as possible into exports. One of the exceptions to the above is Iran, where in certain areas there is extensive residential access to natural gas (e.g. Shiraz and Teheran), and for many years a major intra-city gas pipeline system has been under construction in Iran.

It is difficult to say just how the recent fall in the oil price will influence the economics of Middle East natural gas, but even in that hydrocarbon-rich part of the world there has been a determined effort to replace oil by gas in local consumption. The reason is the superior qualities of oil when it enters into export activities, and by way of contrast the comparative advantage of natural gas when it is used as a refinery fuel and petrochemical feedstock. It takes a great deal of energy to operate the various mechanical and electrical processes in oil refineries, and the ability to use inexpensive gas as a refinery fuel (instead of fuel oil, which has a higher export value) is almost as important for some OPEC countries as the availability of the inexpensive oil that is processed in the refinery. Similarly, natural gas is a more economical petrochemical feedstock than naphtha, which is a refinery output. At present oil prices, naphtha probably costs about 3.5 dollars/MBtu, which can be compared with the 50 cents/MBtu at which Saudi Arabia has fixed the domestic price of gas.

Countries with associated gas have a more complex problem in obtaining inputs for their gas-based industries than those with nonassociated gas. The gas/oil ratio is not uniform in the Middle East, varying from 300 cubic feet of wet gas per barrel of oil to 1,800 cubic feet, and thus there could be times when, to keep an industrial installation or power plant operating at full capacity, more oil would have to be produced than could be readily disposed of. Accordingly, it has been suggested that associated gas lacks the flexibility for applications such as power generation, and that even hydrocarbon rich countries might find it profitable to generate electricity, in nuclear power plants.

NON-OPEC DEVELOPING COUNTRIES

According to the World Bank the costs of natural gas development are lower, and the potential domestic demand much higher, than previously believed; and recent studies by that institution suggest that the cost of producing and transporting gas may be well below the border price of imported petroleum fuels. Whether this is true if the oil price remains in the 15 to 18 dollar/barrel range remains to be seen, but there is no question that the local price of gas in most developing countries makes it an attractive alternative to imported crude and oil products. At the present time only a few of the non-OPEC developing countries are considered to be potentially exporters of crude, although several of the approximately 40 non-OPEC developing countries that have gas also possess a major export potential. In this latter group are Mexico, Thailand and Malaysia.

It is expected that between 70 and 75 per cent of the gas produced in the deveoping countries will be consumed locally. During the next decade these countries will probably use between 50 and 75 per cent of their domestic gas consumption as fuel for electric power and industry, 20 to 40 per cent as a feedstock for fertiliser and petrochemical manufacture, and five to ten per cent in the residential and commercial sectors. This, incidentally, is probably an inefficient consumption pattern, and is almost certainly due to insufficient local distribution systems. At least one major gas producer, Mexico, has decided that priority should be given to the domestic use of gas rather than exports. At present rates of growth the gas that is being produced in Mexico can easily be absorbed locally, and as gas transport facilities are expanded, even more gas can be used. Current gas development projects centre on the elimination of flaring and the conservation of gas reserves. An effort is also being made to rationalise gas prices, which at present fail to express the true scarcity value of gas to consumers in various parts of the country.

Similarly in Egypt, a policy has been established of sanctioning export projects only after a certain amount of gas has been set aside for local consumption. Egypt was one of the first exponents of a strategy designed to provide incentives for gas exploration by foreign companies, while at the same time providing iron-clad safeguards for domestic consumers. According to the *Oil and Gas Journal* (12 April 1982), this involves the creation of a national reserve to meet future domestic needs and the proviso that all reserves discovered by the national oil company become part of the national reserves. Foreign contractors receive compensation for appraisal and exploration costs, with this compensation increasing for larger volumes. The difference between the reserves of the Egyptian government and the reserves located are, in theory, available for export, and gas exports are supervised by both the government and by foreign companies. The share of each foreign company involved in an export project is in proportion to its equity in the total reserves dedicated to the export project.

One country that desires to export more gas is Indonesia, but until a larger export market appears, attention is devoted to increasing the output of gas, and then using this gas in petrochemical and industrial operations. In some studies Bangladesh and Pakistan are pictured as potential exporters, but considering that present plans in these countries call for using at least half of the output of gas for local consumption, this does not appear likely. Bangladesh desires to substitute gas for imported oil, and if this policy is successful, and in addition more sophisticated uses are found for gas, there may not be a great deal of gas left to export. However, if greater reserves do become available, the production costs of gas in both Pakistan and Bangladesh is low enough to make it competitive on world markets — assuming, of course, that it originates in the same type of geological structures that characterise present gas reserves.

The energy medium that has received the most attention in Brazil has been oil, but as greater amounts of gas have been found, plans have been drawn up to rationalise its use. (Both reserves and production of gas have increased by a factor of three in Brazil in the past ten years.) A major problem in that country has been the lack of pipelines, but as more lines are laid, the comparatively low cost of natural gas is causing it to replace both fuel oil and the bottled LPG that is used in the household sector.

In many developing countries, as in Brazil, the cost of imported oil is a major burden on the economy; and in the long run escaping this burden turns on the possibility of finding a substitute for oil that does not have to be imported. Fortunately, there are many situations where low cost natural gas can replace gasoline and diesel fuel, and there are already a number of countries around the world that are using one or more of three processes to produce motor vehicle fuel.

The first of these is the production of LPG by separating out the heavier components of natural gas. LPG has been used as a vehicle fuel for over 50 years, and in the United States some of the vehicles now produced by General Motors function exclusively on LPG. The second process involves converting methane to e.g. methanol — which until recently was primarily used as a fuel for racing cars and as a feedstock in the manufacture of chemicals. Methanol, which can also be produced from coal, is an excellent fuel, and occasionally it is possible to find references to the 'methanol economy'. The third process consists of compressing natural gas into CNG (compressed natural gas) or LNG (liquefied natural gas). Some analysts consider CNG to be an even more satisfactory vehicle fuel than methanol.

At the present time it appears that more emphasis is being put on the use of CNG as a replacement for gasoline and diesel fuels than the other processes mentioned above, but there is a widespread feeling that it is only a matter of time before there is a genuine breakthrough in methanol production and use. This may, however, be wishful thinking, because from time to time there is a fairly large oversupply of methanol, and its producers are prone to overadvertise its good qualities.

In so far as LNG is concerned, we are not talking about the huge capital-intensive installations that are used to liquefy gas for export projects, but small liquefaction plants that can provide LNG suitable for use as a vehicle fuel, or in a form where it can be transported from a well to a remote town or plant and used as LNG or, after processing, as a gas. In this latter example LNG is a substitute for a pipeline. The major operating expense for these LNG plants is associated with the gas used as fuel for the plant: when processing relatively pure gas, almost 20 per cent of the incoming gas is consumed, and this figure increases to at least 30 per cent when the gas must be treated. Another important expense can be the cost of storage.

It does not make a great deal of sense to talk about gas-based development for the developing countries, but the rational use of natural resources such as natural gas might prove to be the key factor in bridging the gap between underdevelopment and development. One of the reasons why the oil price shocks were so damaging to developing countries was that they greatly increased the price of ammonia-based fertiliser. But ammonia manufactured from gas — which can be upgraded to urea — is especially competitive with regard to the ammonia manufactured from naphtha, which in turn is derived from crude oil. The fundamental problem here is getting a large enough domestic market to support a minimum sized ammonia manufacturing plant. This is a problem of organisation, because in most of the Third World the problem is too little rather than too much fertiliser.

This raises an important issue. Fertiliser is important, but so are cooking, electricity generation, cement making, desalinisation, etc. There are only limited reserves of gas available at a given time, or even over a fairly wide span of years. A key problem for countries planning to use gas but faced with a large variety of activities in which gas can be used, is determining the value of gas in terms of these activities. If a full schedule of prices is available, a fairly simple linear programme can provide the desired results; but usually these prices are not available, so both the choice and the technology become a political or bureaucratic process. A typical problem, and one that is characteristic of the entire developing world, has to do with the rapidly increasing demand for electricity. Countries with gas will be tempted to satisfy this demand with power generated in gas turbines, rather than with thermodynamically more efficient steam turbines driven by coal or nuclear reactors. Gas turbines have a low capital cost and, because they are modular, can be constructed more rapidly than steam power plants. But this does not automatically make them suitable for the kinds of applications found in developing countries.

Some of the experience now being gained in exploiting the large gas deposits of the Middle East can probably be employed in other developing countries. The intention of several OPEC countries is to break their dependence on crude oil and oil products by becoming major actors in the market for petrochemicals, and in the long run this means producing — or perhaps just owning the plants for producing — final products, as well as basic and intermediate products. Clearly, many of the potential customers for these final products are in the developing countries, and one way to raise the purchasing power of these potential customers is to take an active part in promoting the economic growth and development of the countries in which they live. This does not mean just giving these countries money — although OPEC's aid programme is in many ways a model of this kind of activity — but the kind of technical assistance that solves rather than creates problems. Just giving countries money is a specialty of the Swedish government, and to my way of thinking has been a disaster for both donors and recipients.

7

The Pacific Region and Canada

This chapter will discuss natural gas in the Pacific region, concentrating on Japan and Australia, and comment on Malaysia and New Zealand. I will also continue the discussion of Canadian gas that was begun in Chapter 4.

Japan occupies the pivotal role in Asia, and as the largest energy importer of the industrial countries, it probably plays the key energy role in the entire world. Unlike most other countries, Japan has a well thought-out energy import and consumption plan designed to take it through the present period of energy uncertainty at minimum cost. First and foremost there is to be a shift away from oil toward a maximum employment of nuclear energy. More coal and gas are also to be used, depending on their availability and on environmental constraints.

Natural gas, in the form of LNG, currently has preference over coal — although this could change at any time. One obvious reason for this choice has to do with environmental considerations: natural gas is relatively 'clean', and its combustion does not involve the large amounts of soot and sulphur oxide emissions which accompany the burning of coal. There are, however, atmospheric environmental pollutants produced in the form of nitrogen oxides, although in the process of liquefaction many impurities associated with nitrogen can be removed. The Japanese are aware that there is a great deal of easily won gas available in most parts of the world. For instance, the Pacific area is rich in gas, and many of the countries possessing this resource are anxious to sell at favourable terms. In many cases gas can be bought for prices well under the equivalent price for oil, which to some extent is due to the manner in which gas contracts are indexed to the oil price. It has also not escaped the attention of the Japanese that when the huge gas supplies of Iran become available to the world market, they may have the same depressive effect on the price of natural gas as the completion of the Siberian-Western European pipeline; and a giant or super-giant gas field could be uncovered in the Pacific region — for many years petroleum experts have believed that one exists in the vicinity of Malaysia. The larger LNG ships can carry a load of energy equivalent to more than 450,000 barrels of oil, and if the right kind of contracts can be signed for LNG, and the worldwide supply of natural gas continues to develop at its present pace, the attractions of LNG can only increase for consumers and potential consumers.

One of the basic issues with which the Japanese are concerned is the programming of gas purchases. If there is going to be a great deal of natural gas available in the world in the next ten years, it would be a mistake to enter into commitments for too much gas too soon. Consequently, the Japanese have not hesitated to abbrogate negotiations where the price offered by the supplier appeared even slightly out of line with what the Japanese interpret as the developing price trend. But they have not hesitated to pay non-competitive prices for energy supplies from some exporters in order to expand their future options. Something that should not be forgotten is that the Japanese have never hesitated to insist on contract renegotiation when they believe that they are facing a long term buyers market.

The history of LNG in Japan is closely associated with the first energy price shock. Japanese imports of LNG increased from 182,000 tons in 1969 to more than 26 million tons in 1984. The first shipments of LNG to Japan were from Alaska (in 1969), but at present five countries deliver LNG to Japan under eight different contracts. Australia has just signed a contract with Japan, and there has been a great deal of talk about deliveries to Japan from Canada and a joint Soviet-Japanese venture involving Sakhalin Island. These will be considered later, but first we examine the present arrangement, listed in the order in which contracts were signed.

1. LNG is still imported from Alaska by Mitsubishi. The import volume is 960,000 tons/year, and the suppliers are Phillips Petroleum and Marathon Oil. The users of this gas are Tokyo Electric Power and Tokyo Gas; the contract is scheduled to run to 1989.
2. Mitsubishi imports 5,140,000 tons/year of LNG from Brunei. The contract was signed in 1972 and is scheduled to run 20 years. The users of this gas are Tokyo Electric Power, Tokyo Gas and Osaka Gas.

3. Matsui imports 2,060,000 tons/year from Abu Dhabi. The contract was signed in 1977, and is scheduled to run 20 years. The gas is used by Tokyo Electric Power.
4. Nissho Iwai Corporation signed two contracts with Indonesia in 1977. The first was with PT Badak for 3,000,000 tons/year; the contract was for 23 years. The users of this gas are Kansai Electric Power, Kyushu Electric Power and Nippon Steel. The other contract, signed at the same time, was with PT Arun for 4,500,000 tons. The users of this gas are Chubu Electric Power and Osaka Gas.
5. Mitsubishi signed a contract with Malaysia in 1983 for 6,000,000 tons of LNG per year. The contract was for 20 years, and the gas will be used by Tokyo Electric Power and Tokyo Gas.
6. Nissho Corporation signed a 20 year contract with PT Badak of Indonesia in September 1983. The contract amount was 3,200,000 tons/year, and the users of the gas are Kansai Electric Power, Chubu Electric Power, Osaka Gas and Toho Gas.
7. Mitsubishi signed a contract in 1984 with PT Arun of Indonesia for 3,300,000 tons per year. The contract runs for 20 years, and the users of the gas are Tokyo Electric Power and Tohoku Electric Power.

Something that can be gleaned from the above is the large amount of gas that is destined for electricity producers. In Japan 76 per cent of LNG consumption is for electric power, 20 per cent for municipal gas and four per cent for industrial fuel. Economic theory stipulates that it is generally uneconomic to carry the base load in electricity generation with gas, which is apparently being done in Japan. Instead, coal or nuclear power should be used. Since the decision makers in Japan can now employ nuclear power more or less as they see fit, and Japanese nuclear installations may be the most efficient in the world, it is probably a mistake to draw the conclusion that the demand for natural gas in Japan will expand in step with the increase in electricity consumption.

In addition to the projects cited above, Japan has been interested in obtaining gas from two other sources. The first was Dome Petroleum of Canada, which was to have sold LNG to three electric companies and two gas companies. The deal broke down, but it is instructive to review some of its details, because it gives an idea of some of the problems that are met by gas traders.

From the Japanese side, initially, no insurmountable difficulties could be detected in the proposed transaction with Dome, assuming that the price was right. In 1981 the Japanese firms that were going to buy the gas concluded contracts for the import of 2.6 million tons annually, and later this was raised to 2.9 million tons when another electric company joined the buying group. On the Canadian side, the sellers had to obtain an export permit from the Canadian National Energy Board (NEB), and also permission from the provincial governments in Alberta and British Columbia. There was apparently no problem with British Columbia — but Alberta — which had exported a great deal of gas to the United States — hesitated. It appears that Alberta felt that the Japanese terms were not particularly appealing, and in addition that a deal with Japan might foreclose more lucrative arrangements with the United States. A Canadian-United States energy conference was held about that time which featured a sustained attempt by the United States representatives to convince the Canadians that the price of their gas was too high, and it should be reduced on existing contracts. In these circumstances the Albertan government came to the conclusion that the Japanese might have something to offer after all.

By this time, Dome's delicate financial condition — which involved, among other things, about six billion Canadian dollars of debt — had caused some anxiety on the Japanese side. Although it was true that the Canadian government was willing to guarantee the transaction because of the exceptionally heavy exposure of five of Canada's major chartered banks, the Japanese understood that unlike diamonds, governments are not forever; and even if they were, they are occasionally inclined to change their minds. Put more directly, Dome might go to the wall before the gas started flowing. In October 1984 the proposed importers served notice that they were losing interest, and in January 1985, Osaka Gas officially withdrew. As things now stand, the project has been set aside by the Japanese — although it could be revived later.

In the matter of the Sakhalin project, it was originally hoped that the Soviet Union and Japan could exploit the gas reserves in the vicinity of Sakhalin Island as a joint venture. Political events and tensions have apparently checkmated that intention; and there seems to be more distrust than ever between these two countries. The main difficulty, to my way of thinking, is that the Japanese have territorial claims in this area, and they feel that if these claims are satisfied, they might be able to exploit a large part of this area by themselves.

The final topic is gas storage in Japan. The International Energy Agency, as part of its programme to make its members more security conscious, has proposed that they greatly increase their storage capacities. Japan has mostly ignored the Agency on this, and to my knowledge has not constructed underground storage facilities. There are LNG storage tanks at LNG-receiving terminals, and together with some other small storage facilities Japan can store about 1.6 million tons of LNG — which is less than ten per cent of their annual consumption. The Agency also wants its members to have more (gas-coal) dual firing capability for both industry and utilities, but strong local opposition has been shown to dual firing power plants on environmental grounds.

NATURAL GAS IN AUSTRALIA

Since early 1985, there has been an oversupply of natural gas in international markets. Large amounts of Algerian gas are moving into Western Europe after an interruption of almost two years; and a number of potential exporters of gas are now fully aware that they cannot expect to experience a sellers market for natural gas in the foreseeable future. Norway, for example, has expressed a willingness to sell large quantities of natural gas to Britain in the 1990s at prices that would have been unthinkable several years ago. Worldwide, the reserves of natural gas are increasing more rapidly than production, and the reserve/production ratio for gas is now more than 58 years (as compared to about 32 for oil). In the United States additions to reserves failed to match production in 1984, but the principal problem in that country at the present time is a general oversupply of all energy materials, to include electricity.

The demand for Australian coal and gas has also been influenced by the supply-demand picture elsewhere, and as a result the Japanese energy market — which has always been important for Australia — is more important than ever. Hong Kong and Singapore also have a major energy deficit, but these markets are miniscule, and do not seem to find Australian supplies especially attractive. China also has a temporary energy deficit, and despite its copious reserves of good quality coal, recently imported Australian coal into Shanghai; but in the long run China should not require imports of energy materials. South Korea and Taiwan are fast-growing economies with only limited domestic supplies of energy, but their energy requirements are not in the same class as Japan's, and Australian producers are not optimistic that they will be interested in large amounts of Australian gas, although they might find Australian coal relatively attractive.

About seven per cent of world natural gas reserves originate in the Far East or the Asia-Pacific region. Australia possesses about one per cent of world reserves, and New Zealand about 0.2 per cent. In terms of those countries possessing the largest reserves of gas (the Soviet Union, Iran and the United States), producers in Asia do not appear to have much to offer, but it seems clear that countries with less than one per cent of world reserves can begin and sustain a profitable export programme. Brunei, for example, which has only about 0.2 per cent of world gas reserves, is one of the main suppliers of LNG to Japan.

Looked at in terms of these figures, Australia appears to have an important role to play in Japan's future consumption of natural gas. It is also clear that gas has become an increasingly significant element in the domestic Australian economy: from almost nothing in 1970, natural gas now supplies nearly eleven per cent of Australia's primary energy consumption. The major gas projects in Australia are located in the Gippsland Basin and the Cooper Basin (which first delivered natural gas in 1969), and the North West Shelf, whose gas is just becoming available in commercial quantities. The North West Shelf is itself a part of the Carnavon Basin, which is the largest basin in Australia, and may contain as much as 500 billion cubic meters of gas, 55 million cubic meters of condensate, 30 million cubic meters of liquefied petroleum gas (LPG), as well as large amounts of oil. The first stage of the North West Shelf project was completed in August 1984, with the delivery by pipeline of gas to the industrial areas of Western Australia. The second stage is the delivery of gas to Japan that was described in the first section of this chapter.

Although the financial rewards from the North West Shelf are considerable, most Australian producers of energy materials now understand that Japan is not shaping up as the bottomless market for Australian gas that it once appeared to be. First, there is too much comparatively inexpensive gas in Asia for Australian gas to appear especially attractive, and the Japanese no longer consider the stability of supply that Australia claims to offer to have a pronounced money value. In addition, estimates of Japanese gas requirements are constantly being scaled down: now that nuclear energy has become politically respectable in Japan, much less gas is going to be required than originally estimated. It is questionable whether Mitsubishi and Mitsui would have invested one billion dollars in the North West Shelf almost ten years ago had they known that the Japanese nuclear sector would obtain *carte blanche* to expand as they see fit, and that world export markets would be flooded with cheap gas.

A large part of the gas imported into Japan is used to generate electrical power, although the ideal fuels for power generation are, of course, coal and nuclear energy — with gas handling peak load requirements. The use of gas for base or intermediate load generation in plants with a thermal efficiency of about 33 per cent thus appears to constitute a drastic departure from conventional economic rationality, but before the politcal go-ahead was given for nuclear power, rigid emission standards and the high cost of suppressing the pollution resulting from burning coal made the use of a great deal of gas almost mandatory. Now, however, opposition to nuclear power has all but disappeared, particularly at the street level, and more efficient pollution suppression equipment is becoming available for coal fired plants. This implies the gradual erosion of one of the most important markets for natural gas in Japan. In 1984 the use of coal in the nine largest power stations in Japan increased by about 15 per cent, presumably at the expense of both oil and gas.

I have already noted the situation in Japan with regard to the use of natural gas in the electricity generating sector; gas has been underutilised in the household sector, though at one time over-utilised in the industrial sector. But where the latter is concerned, conservation and major structural changes in the economy have greatly lowered the aggregate energy intensity of Japanese industry. A developing glut of fuel oil has helped to slow down the introduction or expansion of gas consumption in many applications. This is true, for example, of the household sector, where a distribution system of 248 gas supply companies, without an interlinking pipeline network, does not give the impression of providing much scope for an increased use of gas.

Since all these things have been obvious for a number of years, it has been suggested that Japan has attempted to create a buyers market in LNG by exaggerating — in its official forecasts — its future energy needs. As far as I am concerned, there is no evidence that Japanese economists are more proficient in forecasting than their counterparts elsewhere. The successively lower forecasts of the Ministry of Trade and Industry, and the Japanese Petroleum Association, make it clear that a message is being sent to potential exporters of gas to get their bids in as soon as possible, because there is a definite limit to the amount of gas that the Japanese market can absorb in the next decade. In addition, the Japanese Institute of Energy Economics recently published a study indicating that the demand for LNG may peak in the early 1990s, rather than continue upwards.

Some question has also been raised as to the virtue of seeking additional foreign markets for Australian gas when there are alternative uses in the domestic economy. As in many other parts of the world, attractive options for the consumption of gas exist in the Australian transport sector in the form of fuelling vehicles with compressed natural gas (CNG); transforming the gas to methanol and using the methanol as an input for the production of synthetic petrol, etc. Or it could be put into a trans-continental pipeline and used in household and industrial applications on the eastern and southern coasts of Australia.

There is opposition in Australia to the export, in unprocessed form, of Australian energy and non-fuel mineral resources. This is not a simple dilemma. Domestic energy prices in Australia have tended to be low relative to international levels, and mainstream economic theory provides convincing arguments about the welfare gains that can be realised from the export of gas in this situation. On the other hand, the high level of unemployment in Australia puts a social value on resources that is different from values established by the market. It could be true that if these energy resources were combined with unemployed labour and other unemployed or underemployed resources, they could contribute to an expansion in domestic incomes that would be substantially greater than the revenues obtained by merely exporting natural gas.

Australia is a country where interest is occasionally shown in gas storage schemes, although actual gas storage capacity is still modest. The Australian Gas Light Company (AGL) has suggested that gas could be stored in the Birwood Colliery, which is a disused coal mine near Newcastle. The 40 million cubic meters of gas that could be stored in this installation would constitute four-to-five days' requirements for the Sydney-Wollongong-Newcastle area, with a combined population of about 3.5 million people. It has also been suggested that gas can be stored in the Balmain coal shaft in metropolitan Sydney, as well as in a number of acquifers in the Sydney Basin. Storage in acquifers has also been suggested for Adelaide. In Western Australia, where a large part of the state will be reliant on a single pipeline for a significant part of its energy supplies, storage is a crucial rather than just an important issue.

AUSTRALIAN RESOURCE RENT TAX AND THE GREGORY THESIS

I would like to conclude this part of the exposition by saying something about two extremely controversial topics that have been at the centre of the Australian resources debate for almost the past decade.

The resource rent tax (RRT) is intended to be applied to profits in excess of those corresponding to a threshold rate of return (which is supposed to approximate the rate of return on other projects of equivalent risk): capital expenditure can be carried forward, compounded at the threshold rate, and when cash flow becomes positive (after allowing for both exploration and capital costs) the RRT will be levied on the future cash flow. As presently proposed, it will be levied before company taxation, and will be an allowable deduction against it. This concept can be found in the Alberta profit-sharing royalty on synthetic oil production, the progressive incremental royalty on the Canada Lands (which corresponds to Crown Territory), and to a certain extent in the royalty systems employed in the United Kingdom sector of the North Sea. RRT now applies to Australian fields licensed for production after 1 July 1984, and covers new fields outside both the oilfields of the Bass Strait and the gasfields of the North West Shelf development: these two are specifically excluded. The tax concentrates on the offshore Western Australian fields Harriet and South Pepper, and the Jabiru field that is under the jurisdiction of the Northern Territory. 'Old' petroleum producing operations are also excluded from the RRT.

Highly respected Australian economists have produced analyses designed to certify the virtues or ignominy of this tax. To my way of thinking, they have proved nothing at all. In the form designed by Dr Ross Garnaut, it was supposed to be used in a less developed country (Papua New Guinea) to encourage exploration and production and to confiscate profits. Applied to petroleum it is impossible to determine *a priori* just what it will or will not do; however in those cases in which extensions to old fields lead to large increases in reserves and production, the tax could represent a major tilt in favour of the tax collector and against the producer. On the other hand, there are clearly a great many projects (and the RRT is on projects, and not companies) for which producers will find the RRT superior to other arrangements, and at least one large company in Australia has said so in no uncertain terms. The main problem with the RRT is the problem associated with all taxes: the growing inability of governments to use tax revenues for productive purposes, and in the case of some governments, their total inability to hide this shortcoming.

As for the Gregory Thesis, which at other times and in other places has been called the Dutch Disease, what it concerns is the ill health that can plague an economy in general — and those sectors of an economy in particular that are not participating in a minerals boom — when a minerals-exporting sector begins to expand at a rapid rate. In the United Kingdom, for example, it has been claimed that North Sea oil has been a principal cause of deindustrialisation. What we have here is brilliant economic theory that is inapplicable to the real world — unless politicians, or their advisors, make the mistake of confusing economic theory with physics. Specifically, we are dealing with a situation that could take place, but is by no means inevitable, even if politicians stand on the sidelines without making any attempt to influence developments. The Dutch disease (which was associated with the large natural gas revenues in The Netherlands), created a small group of losers, but increased prosperity for almost everyone else. Norwegian oil has meant about the same thing. One of the principal factors causing the present economic difficulties in The Netherlands is the steady depletion of what once appeared to be a nearly inexhaustible supply of gas; and as in the case of all minerals, this is the real dilemma. Certainly, the point for Australia is not whether to exploit their mineral resources, but how to exploit them optimally.

MALAYSIA AND NEW ZEALAND

There are a number of gas-producing, and potentially gas-producing, nations in Asia, but for years expectations have been that Malaysia would eventually forge to the lead. Gas provides only about one per cent of the country's energy requirements, with hydro providing four per cent and oil almost all the rest; in terms of reserves Malaysia has four times as much gas as oil, and by the turn of the century it seems certain that gas will have replaced oil as the principal source of energy. At present Malaysia has the largest gas reserves of any country in the Far East, and much more will probably be found.

The most important step in the exploitation of Malaysian gas reserves was when the Bintulu LNG plant was commissioned in 1983. This project — the largest ever undertaken in Malaysia —

involved gas being transported via a 36-inch pipeline from Sarawak State (in East Malaysia) to the port of Bintulu, and from there delivered to Japan in a 'train' of 5 LNG ships. The Bintulu plant was constructed by the Japan Gasoline Corporation and M.W. Kellogg Inc of the United States. It is operated by the Malaysian Liquefied Natural Gas Corporation (MLNG), in which shares are held by the state-owned mineral company Petronas (65 per cent), Shell (17.5 per cent), and Mitsubishi (17.5 per cent).

The Bintulu installation is only the first of a number of large projects that are envisaged for Malaysia. The gas for the Bintulu project came from the South China Sea off the eastern states of Sarawak (and Sabah), but there are also very large resources off Trengganu State in Peninsular Malaysia. Much of the Trengganu gas is expected to be used in various domestic industrial operations (e.g. a steel plant and a power station), and some of this gas may eventually be piped to Singapore. It is also expected that some of the Bintulu gas will be used domestically — possibly in an ammonia and urea fertiliser plant, and perhaps also a methanol plant.

As things now stand, Malaysia is not too concerned about speeding up gas exploration and exploitation. The intention is to offer one or two new blocks for exploration each year, and the assumption is that there will be no more incidents of the type experienced with Continental Oil Company of the United States (Conoco) several years ago. It seems that Conoco resented having the outright concession it had obtained from the Malaysian government turned into a production-sharing contract, and withdrew from the agreement. Its exploration area was largely taken over by Petronas' exploration subsidiary, Petronas Carigoli, which has made at least one strike in the vicinity of Conoco's former contract area.

New Zealand's energy policy favours self sufficiency in transportation fuels to overseas markets. The intention is to establish plants to produce CNG, LPG and methanol as transportation fuels, and perhaps an installation which will produce methanol and/or ammonia-urea for export. These plants are to receive natural gas from the Taranaki gas fields.

The New Zealand CNG project is the second largest in the world, after Italy, and plans call for converting 200,000 vehicles to CNG by 1990. The conversion cost is approximately 1,000 New Zealand dollars, but since the price of CNG has been set as half the price of petrol, most motorists are rapidly able to recover the cost. Opponents to CNG point out that engines need reconditioning before conversion, which adds 1,500 dollars to the cost of conversion, but special loans at very low rates of interest are available to cover this expense.

On South Island, where only a third of the New Zealand population lives, the intention is to use LPG brought from the Taranaki fields by tanker. Its cost will be slightly higher than CNG, but still lower than conventional motor fuel. According to recent plans, about half as much LPG is to be used in New Zealand as CNG. The only reason for using LPG is the cost of piping CNG from the North to the South Island.

The final substitute for conventional fuel is scheduled to be motor fuel produced from methanol (which itself is produced by passing natural gas over a zeolyte catalyst). In the New Zealand plant, 4 million cubic metres of gas per day from the Maui field is turned into 4,400 tonnes of methanol, and then processed into 1,600 tonnes of motor fuel. The price of this fuel will be the same as conventional fuel, and it is hoped that eventually it will provide a third of the country's vehicle fuel. Before the price of oil declined, I would have judged this an attractive scheme but if the oil price falls under 15 dollars/barrel, and remains at that level for a long period, I am not so sure.

CANADA

Canadian gas markets are in the process of rapid change. They are under pressure to set their prices according free market criteria. No comment will be made on this transition, although I have made it my business to claim on many occasions that free market pricing will never convince rational investors that it makes sense to put their money into installations having the kind of capital intensity that characterises natural gas systems.

The Canadians are also intent upon exploiting the favourable position they occupy in respect to the gas market in the Northern part of the United States. Although the United States has vast gas reserves (200 trillion cubic feet as compared to 90 trillion cubic feet for Canada), the major American gas fields are in the Southern part of the United States, and while there is an extensive pipeline from the South to urban areas all over the country, a number of Canadian gas fields are ideally located to compete in some United States markets: Canadian gas is only four per cent of United States consumption, but on the West Coast of the United States it is ten per cent.

Beginning in the 1960s, after an extensive review of gas prices and pricing, the National Energy Board (NEB) decided to become more active in determining the prices at which Canadian gas was to be sold. The assumption was that an organisation such as the NEB was more conversant with the overall gas and energy supply and demand picture than private Canadian firms. If, for example, Canadian exporters were being paid less for their gas than American gas producers were receiving, then the Canadian gas supplier was being taken advantage of.

This problem was to be dealt with by manipulating the 'border price of gas'. What this concept meant was that Canadian firms should not charge less than this border price, which was calculated on the basis of the commodity value and the substitution value of gas. There is no point in going into the scientific basis of these concepts, because they have none: clearly, a uniform border price does not make sense given the length of the United States-Canadian border and the heterogeneous nature of gas markets on both sides of the border. Ostensibly, gas was not to be sold under full cost, and particular cognisance was to be given to both the price of substitute fuels and the world price of energy. Accordingly, as the world price of oil increased, the border price of gas was steadily adjusted up. It was 1 dollar/MBtu on 1 January 1975; 1–2.10 dollars/MBtu in 1976; 3.45 dollars/MBtu in 1979; and 4.47 dollars/ MBtu in 1980. In April 1981 it was increased to 4.94 dollars/MBtu, and a further increase was scheduled for October which did not take place, because the demand for gas was growing weaker. In April 1983 the price was adjusted down to 4.40 dollars/MBtu, and with the market in full decline it fell to 3.40 dollars/MBtu in July for certain categories of gas.

If we try to fathom what the Canadian government is attempting to do, what we see is that it has a perfectly sensible ambition to introduce and implement a system of price discrimination in foreign markets, from whose vagaries Canadians would be shielded as much as possible. Ideally Canadian gas revenues would be maximised, and the same is true of government revenues that were associated with the sale of natural gas; and conservation — which has become a sensitive issue for many voters — would also be promoted. By way of bringing all this about, the alleged quantitative skills that economists possess would be used by the government to make calculations that private firms — supposedly — do not make, or have no interest in making (or even both). With these calculations in hand steps would be taken to ensure that gas buyers in the United States or elsewhere did not escape giving Canadian sellers every cent of their due. A more technical discussion of this issue can be found in an important contribution by Watkins and Waverman (1985).

APPENDIX: NATURAL GAS SHIPPING

A key item in a typical LNG chain is shipping. Tables 7.1 and 7.2 provide data about LNG shipping and, for comparative purposes, information about oil carriers.

Table 7.1: LNG shipping (1986)

Status (LNG carriers)	Number of vessels	Capacity (cubic meters)
Long term contracted		
Operating	41	4,254,220
Laid up and awaiting project start-up or resumption	10	1,236,200
Short term contracted and spot chartered	5	190,700
Uncommitted or laid up	13	1,355,400
Total	69	7,036,520

Source: International Energy Agency.

Table 7.2: Oil tankers and combined carriers (1984, 1985)

Oil tankers and combined carriers size: (deadweight tons — dwt)	1984 Ships	1984 Million dwt	1985 Ships	1985 Million dwt
10,000– 25,000	674	13.2	645	11.4
25,000– 80,000	1,309	59.2	1,275	58.2
80,000–160,000	675	76.3	645	72.9
160,000–320,000	496	122.1	430	106.2
320,000–	71	27.9	64	25.2

Source: Australian Institute of Energy, annual report (1985).

8

The Western European Natural Gas Economy

If natural gas is going to be in oversupply for at least the next five or six years, which is possible, Western Europe may be the part of the world that will benefit most. One reason is that Western Europe is the world's most important market for imported natural gas. This chapter will present a comprehensive survey of the Western European natural gas economy, with special emphasis on Norway.

INTRODUCTORY SURVEY

In 1984 natural gas consumption in Western Europe was 212 billion cubic meters (= 212 Gcm) of natural gas, or 188 million tons of oil equivalent (= 188 Mtoe). This was approximately 15 per cent of Western Europe's primary energy demand, compared to two per cent in 1960. Ninety per cent of the proved gas reserves of Western Europe are in the North Sea or the offshore gas fields of The Netherlands, and about half of the natural gas consumed in Western Europe reaches consumers via international trade, supplied by The Netherlands, Norway, Algeria, Libya and the Soviet Union. (Primary energy is energy obtained from the direct burning of oil, coal, gas, etc. as well as electricity having a hydro or nuclear origin. Electricity obtained from e.g. the burning of coal is a secondary energy source.)

Gas is subject to heavy competition from other energy media in Western Europe, but there is still considerable scope for further expansion. Expectations are that the demand for gas will expand to at least 300 Gcm by the year 2000. There is an extensive pipeline system or gas grid in Western Europe that is ideal for shifting gas from one country to another should the need arise, and there is also strong inter-country cooperation, with the International Energy Agency (IEA) generally — but not always — functioning as an advisory body. The major transmission pipelines are owned by national firms such as SNAM (Italy), Ruhrgas (Germany), Gaz de France and ÖMU (Austria). Producing and processing companies such as Shell and Esso have interests in trunk pipelines in West Germany. This pipeline system extends from the North Sea to the Mediterranean, and from the borders of Eastern Europe to the Atlantic Ocean. The grid is thinnest in Scandinavia, but even in these countries it is being extended.

An important characteristic of the European gas sector is that only a relatively few companies are involved in the production and transmission of gas. The market is almost entirely dominated by large national companies that are complemented by multinationals such as BP, Conoco, Esso, Mobil, Phillips and Shell. Where gas is purchased from domestic producers for transmission and reselling, there is usually one dominant buyer — although in Germany long term contracts are employed which specify the share of new domestic supplies going to the various pipeline companies. For international transactions, multinational consortia are formed, and these not only carry on negotiations with actual and potential suppliers, but redistribute gas supplies among consortium members according to evolving demand. Table 8.1 shows the 1984 demand for gas by the most important consuming countries, its use by sector and the share of gas in primary energy demand for these countries.

Traditionally countries with large domestic supplies of gas tend to be lavish in its consumption. The Netherlands, for instance, has made a virtue out of using gas for power generation, although

Table 8.1: Natural gas consumption and share of gas in six Western European countries (1984)

Country	Consumption (Gcm)	Share of gas in each sector			Share of gas in primary energy %
		Household %	Industry %	Power %	
Belgium	8.7	33	22	7	17
France	24.0	24	19	0	12
Germany	46.0	23	21	7	16
Italy	31.6	33	25	15	20
Netherlands	35.1	75	40	65	47
United Kingdom	50.8	50	30	1	22
Others	12.2				
Total	212.0				

Source: Calculated from International Energy Authority, BP and Shell statistical summaries.

economic theory indicates that only a modest amount should be used in this application — mostly for peaking purposes. Overall, however, there is a tendency in Western Europe to decrease the use of gas in electrical power generation. In Western Europe, as in the other OECD countries, a change is underway from high energy intensity heavy industries to low energy intensity high tech or low tech activities. Energy intensity (= energy consumption per unit of gross domestic product) has fallen by 21 per cent in the OECD between 1973 and 1983, while oil and gas intensities fell 32 and 23 per cent respectively.

Belgium. Belgium is an energy-poor country, although it has some coal and the voters' permission to use the nuclear option. Gas is being imported from The Netherlands, Norway, and in LNG form from Algeria. The industrial demand for gas has been declining rapidly the past few years due to the fall in economic activity, and also because gas is being withdrawn from power generation.

France. France makes a point of practising widespread diversification where its supplies of natural gas are concerned. It buys from The Netherlands, Norway and the Soviet Union, and takes a large amount of LNG from Algeria. Someone in France has called the price of Algerian gas theological, because initially it was higher than the average price of gas being traded elsewhere in Europe. Political would have been a better word. Domestic supplies of gas are depleting rapidly, and the French coal industry is uneconomical. France's energy future will largely be based on nuclear energy and imported gas; if the French decision makers get their way, it will mostly be the former. Sixty five per cent of French electricity requirements are now covered by nuclear power, compared with e.g. 59 per cent in Taiwan, 58 per cent in Belgium, 40 per cent in Switzerland, 32 per cent in Bulgaria, 31 per cent in West Germany and 25 per cent in Japan. The hope of the energy bureaucracy is that by the year 2000 at least 75 per cent of electricity consumption will be nuclear-based, and nuclear reactors will provide 40 per cent of all energy. (The United States has the most reactors but only 14 per cent of its electricity is generated in nuclear installations.)

West Germany. West Germany produced 20 per cent of its gas requirements in 1984 and because exploration has been reasonably successful in the past few years, the present level of production can probably be held for the remainder of this century. The cost of

German gas is increasing, however, because it is necessary to drill deeper, and because the gas fields are fairly small by international standards there are virtually no economies of scale. Gas is used in power stations only where special contractual obligations still exist. Outside of this fairly small amount, gas has steadily been withdrawn from baseload electricity generation over the last decade. West Germany is a large importer of gas from The Netherlands (from which is obtains about 30 per cent of its supplies), Norway (14 per cent) and from the Soviet Union (which is expected to provide 30 per cent of supplies by 1990). West Germany has also expressed interest in obtaining LNG, but has not purchased any to date — although on several occasions preliminary agreements were reached with Algeria.

Italy. There were originally large supplies of gas in the Po Valley, although these are depleting rapidly. Italy is a large importer of gas, obtaining more than half its requirements from The Netherlands, the Soviet Union and by the TransMed pipeline from Algeria. During the past year or so Italy has experienced a considerable excess supply of gas (due, in part, to take-or-pay stipulations in the Algerian contract), and as a result is using gas in such non-recommended applications as baseload power generation. The use of gas (and oil) in baseload power generation seems likely to continue beyond 1990.

Netherlands. Because of the super-giant Groningen gas field, The Netherlands was able to become the largest gas exporter and one of the most intensive gas users in the world. By intensive I mean that gas is used in the domestic economy to a greater extent than in any country in the industrial world, and this includes a large role for gas in non-premium applications. Given the chance, however, gas is to be phased out of its role in baseload electricity generation. Because of Groningen gas, The Netherlands was able to construct a comprehensive social welfare system that eventually became grotesque in some of its ramifications; but now that the gas is depleting, some of the cornerstones of that system will have to be dismantled.

United Kingdom. The large gas supplies in the British sector of the North Sea are depleting rapidly, although the present British government feels that if the domestic price of gas were higher, more would be found. Accordingly, one of those 'deals of the century' — this time between Britain and Norway for gas from the Sleipner field —

was cancelled. Britain already imports a large amount of gas from the Norwegian sector of the North Sea, but up to now exports of gas have not been permitted. Increased privatisation, which is the policy of the present government, could increase supplies, but I have my doubts. The latest revision of UK North Sea Reserves reversed the previous upwards trend and cut the estimate of UK gas reserves by the equivalent of a year's consumption.

Denmark. Like The Netherlands, Denmark is a country with severe economic problems that, until recently, were patched up with a welfare system that is rapidly growing more unwieldy — and unjust. The presence of gas in the Danish sector of the North Sea is a considerable windfall for a country that, until recently, appeared to be ready for the title of Sick Man of Europe. Present intentions are to market a considerable amount of this gas in West Germany and perhaps Sweden. Domestically gas will be used in power stations, because nuclear energy is taboo. More oil may also be used now that its price is falling.

Spain. Spain imports LNG from Libya and Algeria, although the Libyan contract does not have many years to run. Spain, like France, has paid prices for LNG that are excessive in terms of the average European price for gas, but this situation is changing due to excess supply on the European gas market. For some years now a pipeline between Algeria and Spain has been discussed, but given the developing supply situation with LNG, a pipeline probably does not make much economic sense, now or in the near future.

Finland. Finland obtains the major part of its energy supplies from the Soviet Union. Since the Soviet Union pays for a large part of its imports from Finland with energy materials, Finland often finds itself with a surplus of such things as oil — which it attempts to sell to other countries. Finland has a small nuclear sector that is among the most efficient in the world.

Sweden. A decision has apparently been taken by the Swedish energy bureaucracy to replace nuclear equipment by gas in the generation of electricity. Nuclear energy cannot be 'replaced' by gas in the sense that the Swedish minister of energy — and her advisors — intend without considerable economic losses. The foundation is being laid for an economic disaster. The gas to be used instead of nuclear energy would come from the Soviet Union (via Finland) or, most likely, from Norway — where the price would be sky high. Towards the end of the 1990s, even with nuclear energy, it has been calculated that Sweden would be using about 2 Gcm of gas, which amounts to about five per cent of its energy consumption. For environmental reasons coal has lost its popularity.

One of the major factors in determining the future of gas in Western Europe is the price of oil, and this works in two ways. If the price of crude oil falls, so does the price of fuel oil, which is a competitor of gas in several key applications. At the same time a fall in the oil price pulls down the price of all energy materials. First there is the 'indexing' effect, which results from the fact that a great deal of the gas being bought in Europe is indexed to the price of crude oil — or, more exactly, this gas is indexed to a bundle of crude oils and oil products. Thus the price of pipeline gas from the Soviet Union has been reduced from 4.65–4.80 dollars/MBtu to under four dollars; and the price of LNG from Algeria has also fallen. Everything considered, if the price of oil is due to remain under 15 dollars/barrel it may mean the use of more gas rather than a return to higher demand for oil — assuming, of course, that there is no drastic decrease in the excess supply of gas that characterises the market at present, and that the price of oil does not fall under twelve dollars/barrel, and stay there for a fairly long period. Something else that has to be considered here is the effect of a low oil price on economic activity in general: if the price of oil were to remain under 18 to 20 dollars/barrel for an extended period, economic growth in the industrial world would be boosted to 3 to 3.5 per cent/year, which would result in a strong increase in demand for all energy materials. Since gas is highly favoured on the basis of price and environmental suitability, a large absolute and relative increase in the demand for gas should be forthcoming.

Some of us believe that the price of oil cannot be expected to remain under 15 dollars/barrel indefinitely. We expect a definitive upwards trend to begin in the 1990s. There are other analysts who believe that the strong economic growth brought about by falling gas and oil prices will eventually cause the gas market to tighten, perhaps very rapidly, and the gas price to rise. In a paper presented at the 1985 international conference of the IAEE, Weale and Pariente-David (1985) of DRI's European Energy Service forecast that real gas prices would fall in Europe during a limited period (which apparently means most of the remainder of this decade), after which they would rise to the present level before the turn of the century. By the year 2000 they would be rising very rapidly. The DRI oil price story is roughly the same.

There is one distinct environment in which this scenario could turn out to be true. If restrictions are put on the import of Soviet gas, and if nuclear energy does not take over the major role in electricity generation, and more gas from the Gulf does not enter the world market, then continued expansion in the demand for natural gas at the present rate will mean that Western Europe will have to use an increasing amount of high-cost Norwegian supplies; and as more of these high-cost (i.e. high price) supplies come onto the market, all prices will be adjusted up. This is how markets work, and so the opinion here is that the energy consumers of Western Europe should leave no stone unturned to ensure that their political and bureaucratic masters are aware of the consequences of not taking steps to safeguard future energy supplies. They have only a few years left, because it will take between five and ten years for new supplies to come on stream after contracts are signed.

The most important use of gas in Europe is in the generation of heat for both space heating and process heat, with the latter market dominating industrial gas use. Gas plays a fairly minor role as a feedstock for chemical processes, although in the industrial sector more gas is used by the chemical industry than elsewhere. The fairly rapid introduction of new condensation boilers (which are more efficient than oil-fired boilers) has helped to increase the demand for gas, although the general slowing down of industrial growth in Europe has depressed the consumption of gas, *ceteris paribus*.

TRADE, PRICING AND STORAGE

In the previous section it was pointed out that the demand for natural gas was scheduled to grow in Western Europe. But since most of the countries in Western Europe are without domestic gas (and even The Netherland's production of gas is falling, and will fall faster towards the end of this century), imports are going to be more important than ever. The problem is that on the basis of existing contracts, the supply of gas could become inadequate toward the end of the 1990s. New supplies are going to be needed about that time if the price of gas is not to accelerate upwards, and investments associated with these supplies should be made in the near future.

As things now stand much of the gas will have to be supplied by Norway and the Soviet Union. Norway, of course, would like to be given the opportunity of shouldering this burden, but many of us who take this matter seriously (i.e. understand the importance of energy) want to see as much Soviet gas as possible purchased: five of the ten largest gas fields in the world are in the Soviet Union (to include Urengoi, the largest), while Troll — the largest Norwegian field — is only tenth largest. Moreover, despite what one hears from Norway in this matter, Soviet gas is and always has been relatively inexpensive.

In the long run pipeline gas may not be enough; even if it is sufficient in quantity terms, as the gap opens between gas already contracted for and the likely demand, consumers should have made provision to close that gap with a considerable amount of LNG. At the present time, 13 per cent of gas consumed in the world enters into international trade, of which three per cent is LNG, but it is expected that this situation will change, with LNG progressively taking a larger percentage of gas trade. Since 1970 the amount of LNG entering into world trade has grown from 2.7 Gcm to 48.2 Gcm, with Japan and France showing the greatest increases in consumption. One of the most important factors for maintaining the world gas market in its present favourable state for consumers would be a stagnation in gas demand by Japan, which is possible but not likely, and the expansion of gas supplies in the Pacific Basin, which is both likely and possible. In these circumstances Western Europe could become a more attractive market for Abu Dhabi and Qatar, which at present expect Japan to purchase a great deal of their gas. Figure 8.1 shows actual and potential LNG projects involving the Atlantic Basin.

Figure 8.1 is more or less self explanatory, but one point can be made with respect to Algeria. Although there is no gas trade between Algeria and the United States (due to previous contracts being suspended), it seems likely that the United States will receive Algerian gas in the not too distant future. As far as I know the amounts have not been settled, but to begin with it seems unlikely that they will be as large as the amounts in the previous contracts.

One of the agonies suffered by many students of economics is the realisation — usually later rather than earlier — that the functioning of real world markets is often completely different from the markets described in textbooks. Anyone setting out to understand the natural gas market eventually finds this out.

In some uses gas commands a high price. One of these is cooking and water heating, and this is so because the alternative to gas is comparatively highly-priced electricity: electricity delivered to

Figure 8.1: LNG projects in the Atlantic Basin

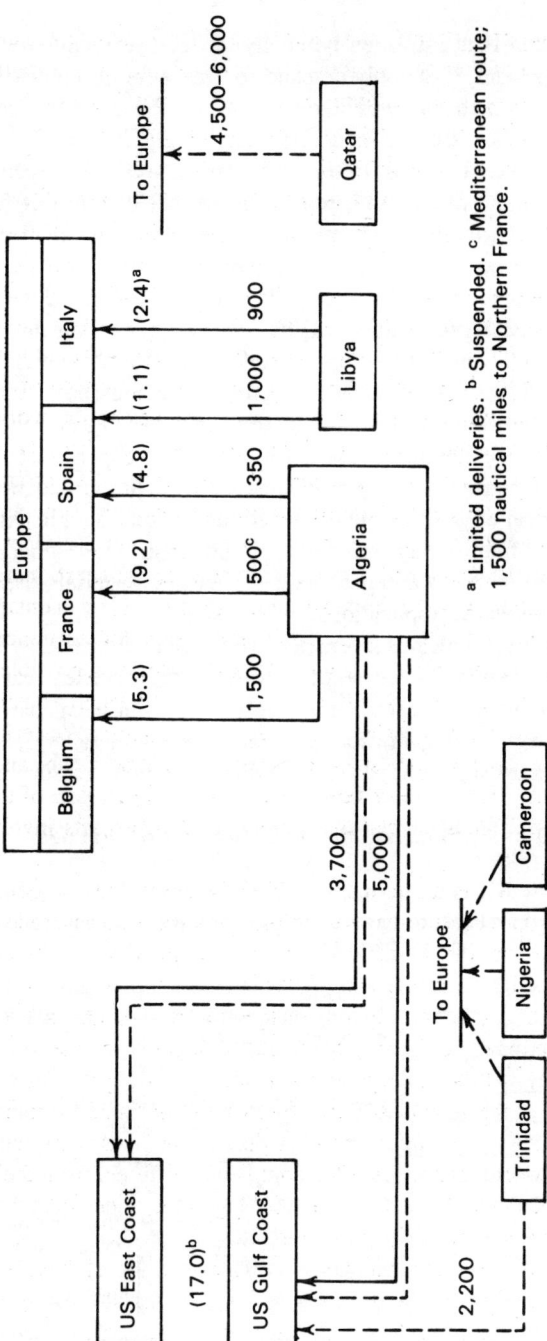

[a] Limited deliveries. [b] Suspended. [c] Mediterranean route; 1,500 nautical miles to Northern France.

Note: All distances in nautical miles, and amounts of gas are plateau volumes in billions of cubic meters per year. The dashed line means possible or planned transactions. Amounts of gas are shown in parenthesis, and distances from seller to buyer.

houses and flats can often be costly, at least in relative terms. On the other hand, if the only demand for gas is for steam raising, then the price which gas would receive would be comparatively low, because steam can be generated employing inexpensive nuclear energy and coal (at least in some countries), and rational people may not be willing to pay high prices for gas to perform this function.

Naturally, there are exceptions. If there are severe restrictions on SO_x and NO_x emissions, as in Japan, a premium is paid for low pollution fuels in such things as electricity generation. Figure 8.2 is a schematic representation of the value of gas in various applications, and the thing to be appreciated here is that if gas was confined to only its premium uses (in, for example, the household sector) then often there would only be a relatively small volume demanded, and because of economies of scale it might be too expensive to provide this small amount at competitive prices. To reduce production costs, a great deal of gas must be produced, and a considerable part of this gas could find its way into non-premium applications where it must be priced so as to compete with the energy materials used in so-called low-value applications.

Figure 8.2 indicates it is expensive to use nuclear energy for peak load applications — i.e. expensive to increase nuclear capacity in order to just use it a short time every day — but inexpensive to use it for steam raising in order to generate the electricity baseload.

When the supply of gas is expanding, and there is a surplus on the

Figure 8.2: Hypothetical demand curve for natural gas

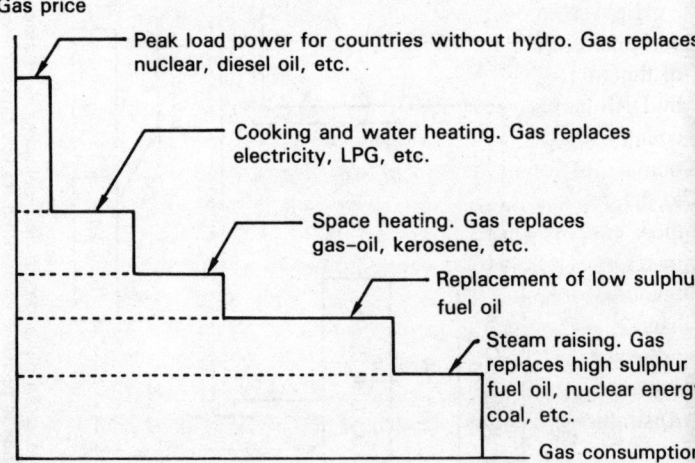

market, it is common practice to reduce prices in order to enhance market penetration and to maximise profits in the long run. Once customers have invested in the durable equipment required to use gas, they are often hesitant to stop or reduce their demand for gas once the price rises. This is one of the reasons why some economists argue against the price of gas to final consumers being set on a free market. On the other hand, when the supply of gas from producers is contracting, distributors often attempt to eliminate sales to non-premium customers as much as possible. In these circumstances, with excess capacity in transmission and distribution pipelines, the regulatory authorities in the United States, for example, have permitted pipeline companies to top off their supplies with high priced LNG, and 'roll-it-in' with low priced domestic gas.

The setting of the price of gas is done through the process of negotiation between buyers and sellers of gas. During the last decade, when the demand for gas was increasing fairly rapidly, the general feeling among buyers of gas was that they could afford to pay prices of gas (measured in dollars/MBtu) somewhere in the vicinity of the price of oil products that were going to be used for similar applications. This almost always meant substantial profits for the sellers of gas, but usually buyers did not feel that they had been taken advantage of. At the same time, however, some exporters were insisting on being paid prices that were equivalent to the price they were receiving for their oil at the point of loading — that is to say, FOB prices. Algeria, for example, wanted El Paso Gas (USA) and Gaz de France to pay a price for LNG equal to the FOB heating value prices of high quality Algerian export crudes. This would have made the burner tip price of El Paso's crude in the United States almost five times the price of domestic gas delivered to electric plants, and at least twice the price of low sulphur domestic fuel oil in the United States. Since El Paso was unwilling to pay more than 4.30 dollars/MBtu, deliveries were cut off in April 1980, and the contract between Sonatrach (of Algeria) and El Paso Gas was suspended. At the time of suspension it was the world's largest single LNG transaction, amounting to 10 Gcm/year. Eventually all contracts and proposed contracts between Algeria and the United States were suspended or dropped, but with Algeria searching for markets in order to use its excess LNG capacity, it is likely that new transactions will eventually be agreed on. Similarly, deliveries to Gaz de France were suspended in 1979 but resumed in 1980, and in February 1982 contract renegotiations were concluded, with the base price of the gas to be traded settled at 5.10 dollars/MBtu. This price was to be indexed to a basket of crudes selected by both buyers and sellers.

Indexing is not complicated but it lacks a 'scientific' basis in that it is impossible to provide a scientific background for the choice of indexing formulae. The base price for United States imports of Mexican natural gas was 3.625 dollars/MBtu, and this was calculated using a 20:80 ratio of the average price of No. 2 (heating) and No. 6 (fuel) oil in a representative sampling of American cities. Note that fuel and heating oil would be considered competitors to natural gas in many applications. In the Algeria-Gaz de France contract, eight OPEC crude oils are used in the indexing formula, with these crude oils given equal weight. In the West German-Soviet (Yamal) contract, the index is composed of 20 per cent crude oil, 40 per cent gas-oil, and 40 per cent low-sulphur fuel oil. The items entering into the index are negotiated in the same manner as the base price. Sometimes it is convenient to write out an indexing formula, as with the West Germany-Norwegian (Ekofisk II) agreement. This is:

$$P = 16.8 \left(0.5 \times \frac{A}{200} + 0.20 \times \frac{B}{275} + 0.30 \times \frac{C}{170}\right)$$

This price is in DM per billion calories.

In this formula A is the price of heavy fuel oil (up to one per cent sulphur content) for German gas customers at Düsseldorf, etc, plus 60 per cent of the price difference between a standard fuel oil and low sulphur fuel oil. B is the price of gas-oil, and C is the FOB price of heavy fuel oil (up to one per cent sulphur content) in Rotterdam. There would also be subsidiary agreements associated with this price, and particularly the base price of 16.80 DM/Gcal. An example of a situation where specific attention is paid to the minimum price of gas is the formula settled on for a proposed deal between Algeria and the United States, which I write in the following form.

$$P = \text{maximum} \left[\frac{P_0}{2} \left(\frac{F_2}{F_0} + \frac{F_6}{F_{10}}\right), MP \right]$$

MP is the minimum price, which was to be calculated monthly on a base of 1.30 dollars/MBtu. The routine for this calculation is given in Davis (1984). P is the invoice price, P_0 the base price (= 1.30 dollars/MBtu on 1 July, 1975), F_2 the price of No. 2 fuel oil in New York Harbour, F_0 = 12.642 dollars, F_6 the price of No. 6

fuel oil (low pour, 30 per cent maximum sulphur, delivered to New York Harbour), and $F_{10} = 13.505$.

In the discussion above — which to a considerable extent was concerned with gas negotiated for in the early 1980s — the price range that almost everybody seemed to think appropriate for the mid-1980s was 4.5–5.5 dollars/MBtu. As a result of the sudden collapse in the price of oil in the beginning of 1986, Western European gas prices appear to be headed for 3.5 dollars/MBtu; and in those pricing formulae where crude oil is the dominant element, they should go even lower. Something else that is important in the context of this discussion is the ownership of the crude that enters into the indexing formulae. Algeria greatly desires to sell more LNG, and on the basis of 'official' OPEC prices, the pricing formulae on which it is selling most of its present gas gives an average price of about 3.80 dollars/MBtu. If the crudes on the Algerian formulae were priced employing Rotterdam spot quotations, the Algerians would be selling gas for under 2.50 dollars/MBtu. Similarly, the delivered (i.e. border) price of Soviet gas to Italy is now 3.1 dollars/MBtu, as compared to 4.35 dollars/MBtu for Algerian pipeline gas. Among other things this means that it will be difficult for the Algerians to convince the Italians to sign any contracts that are similar to those under which they now purchase Algerian gas.

Two more items should be considered before completing this section. The first is the concept of border prices, which was raised earlier in this book in the context of United States-Canadian gas trade. By agreeing on a border price, it is possible to eliminate the distinction between FOB and CIF prices, which often prove to be a time consuming bone of contention in negotiations between buyers and sellers of gas. There is also the matter of minimum billing or take-or-pay clauses that are intended to safeguard the investments of gas producers and pipeline owners. These arrangements require that consumers of gas take a certain amount in a particular contract, or pay for this gas. The Algeria-Italy transaction is one in which the Italians have found themselves with a surplus of gas because there is no way to avoid taking the gas without being penalised.

It is possible to discern a kind of optimal behaviour for producers. To begin with, they should not be too concerned with delays in signing contracts for future deliveries of gas. If consumers do not take steps in the next two or three years to sign contracts for the gas they will need in the mid 1990s, it seems likely that there will be excess demand for natural gas by the turn of the century, and gas sellers may find themselves in the position to reap the high profits that many of them have been dreaming about during the past decade. They might also consider re-enlisting the assistance of the United States government: the attempted boycott of Soviet gas by President Reagan, and the bogus security issue it raised, will probably cost Western European gas consumers hundreds of millions of dollars — or more — before the end of the first decade of the next century, since it will keep a great deal of fairly inexpensive Soviet gas out of the Western European energy market.

STORAGE

The matter of storing gas was referred to in the previous chapter in the section on Australia, where it was noted that one of the main reasons for storage in that country is fear of an accident that could interrupt the supply of gas to one of the large gas-using Australian cities. The main issue in Western Europe is security against an external interruption of gas supplies: Algeria, dissatisfied with prices, refused to fill the TransMed pipeline for two years, and changed its mind only when new supplies of Soviet gas began reaching Italy; while many politicians in Western Europe consider the Soviet Union to be an unreliable supplier.

Since gas storage capacity in the United States amounts to more than 25 per cent of annual consumption, the IEA feels that Western Europe storage capacity could easily be boosted by a large amount. Storage capacity was doubled between 1970 and 1978, but because gas consumption also increased, storage capacity as a percentage of annual consumption actually fell. Table 8.2 shows the storage situation for the OECD at the end of 1983. Capacity is in Mcm.

It is considered important not just to increase storage capacity, but to ensure that gas stored in one country can be routed to another. Once this can be done on a large scale the probability of having to employ crisis management in the Western European gas economy will be greatly reduced, since a gas exporter that was tempted to cut off gas deliveries to a particular country would realise that it was a wasted gesture.

The OECD also apparently believes that the flexibility of the European Gas Grid would be greatly enhanced if the British Isles were linked with continental Europe, and if the transmission capacity between Northern Italy and Spain to France were increased. At present the United Kingdom receives some gas from Norway, and a pipeline between Britain and The Netherlands has

Table 8.2: Gas storage facilities in OECD countries, including facilities under construction (UC), (Mcm), (1983)

Country	Facility	Capacity
Belgium	LNG	70
	Former mine	80
	UC	740
France	Acquifers	7,200
	Salt Dome	120
	UC	3,620
West Germany	LNG	31
	Acquifers	565
	Depleted fields	2,000
	Salt cavern	856
	UC	2,298
Italy	Depleted fields	9,360
	UC	4,270
Japan	LNG	2,230
United Kingdom	Salt cavern	187
United States	Underground storage	212,000
	LNG	3,430
TOTAL OECD		215,430

been proposed on various occasions by continental gas users who want access to low priced United Kingdom supplies, as well as by politicians and civil servants in the United Kingdom who believe that later in the century it might be desirable to have access to gas supplies from The Netherlands or elsewhere. The elsewhere is, of course, the Soviet Union, and the lack of popularity of this proposal in Washington and with the present government in London, was one of the reasons why the British Gas Corporation entered into negotiations for 30 billion dollars of Norwegian gas. This will be considered later in the chapter.

In the discussion above it was mentioned that storage is resorted to for security reasons and for fear of an accident, but it is also useful to smooth out production over the annual cycle: if gas is stored when demand is low, it can be released from storage when demand is high (e.g. during peak periods). The United Kingdom now has a system in which the Rough gas field absorbs gas from other fields during much of the year, and especially the summer months when demand is low; and then releases this gas during the coldest 80 or 90 days of the year when demand passes a certain level. This scheme, which cost almost a billion dollars, was chosen as an alternative to drilling more wells in North Sea fields: had more wells been drilled, they would have remained idle over most of the year, and this would not have justified the high cost of drilling these wells and connecting them to the pipeline system. Present plans are to combine the Rough field output with that of the Morecambe Bay field during peak load periods. These two fields can supply a quarter of the United Kingdom's gas demand during peak periods.

Another device for meeting peak demand would have been to store LNG in large pressure vessels (i.e. tanks). To get the same effect as the Rough field would have required 80 of these tanks, each with facilities for liquefying and deliquefying the gas, and in addition located well away from populated areas for safety reasons. (Some of the worst industrial accidents of this century have been caused by exploding LNG.) The cost of each of these LNG complexes would be between 42 and 56 million dollars (at today's exchange rates). Thus, given the presence of storage capacity in depleting gas fields, the use of LNG tanks represents a sub-optimal solution — at least for the United Kingdom.

Mechanically the Rough field system functions by pushing gas back into the sandstone structure of the field under great pressure. It has long been appreciated that depleted gas fields provide excellent storage facilities, and more than once it has been suggested that the Groningen field would be the premium storage area for Western Europe. If this is true, and if there really is a security problem due to importing large amounts of gas, then the growth in storage space resulting from depleting Western European gas fields, together with an increase in the facilities for receiving LNG from sources around the world, and a marginal extension of the European Gas Grid, should soon reduce that problem to a minimum.

UNITED KINGDOM, FRANCE AND WEST GERMANY

United Kingdom

The British Gas Corporation (BGC) is the world's largest gas corporation, functioning both as a monopoly and a monopsony. As a monopoly it sells gas to United Kingdom consumers — and at uniform prices that are not characteristic of monopoly prices; while as a monopsony, in control of United Kingdom transmission pipelines, it is charged with buying natural gas from producers in, offshore and outside the United Kingdom. Some major changes may come about in the behaviour of this organisation as a result of the Oil and Gas Enterprise Act of 1982, which forces the BGC to make its distribution network available to third party carriers, and thus e.g. enables gas from the North Sea to be sold to inland United Kingdom buyers by North Sea producers. This new arrangement, which was devised by the present United Kingdom government, is supposed to increase consumers' efficiency and welfare where, by efficiency, it is meant that the present uniform price schedule of the BGC causes some consumers to take too much gas and others too little. The opinion here is that in the short run United Kingdom consumers are best served by the present arrangement (and perhaps in the long run too), and that the changeover costs associated with another system of gas pricing would outweigh any theoretical advantages that are supposed to accompany price differentiation.

In its original form the BGC was called the Gas Council, and it exercised a supervisory function over 12 area boards. The Gas Council dated from 1948 (and the Gas Act) and the first nationalisation of the post-war Attlee Labour government in 1949. In 1972 the National Gas Act set up the BGC, and the Area Boards became departments within the BGC, with increased accountability to the central organisation. The first international shipment of LNG was arranged by the Gas Council (from Lake Charles Louisiana to Canvey Island in 1959), and a short time later the Gas Council was a signatory to the first long term LNG contract (with Algeria).

Interest in offshore possibilities began with the first finds in the Groningen gas province (at Schloctern, The Netherlands) in 1958. Exploratory activities began in the United Kingdom North Sea in 1962, and the first licensing round took place in 1964. The following year the first gas strike was made by British Petroleum (the West Sole field), and the largest of the British North Sea fields, the Leman field, was discovered in 1966 by Shell-Esso. (The Frigg field is larger, but a part of this field is in the Norwegian sector of the North Sea.)

The monopolistic and monopsonistic character of the present BGC was established in 1965, when the Gas Council was given the right to purchase North Sea gas destined for Britain, and to distribute it to the Area Boards for sale to final consumers. Gas prices, at least to begin with, were extremely low — so much so as to bring charges of dumping against the Gas Council. The reason for the low prices, of course, was to accelerate market penetration for the large amounts of gas that were due to be landed. Within a fairly short time the Gas Council was able to pronounce its efforts a success, because not only was their customer base expanding rapidly, but they made a profit. On the other hand it was claimed, and to some extent proved by the Price Commission of 1979, that the marginal cost of North Sea gas was more than the price being charged to consumers, and accordingly there were solid economic grounds for an increase in the price of this gas. (Among other things, an excessively low price for gas would cause consumers to make investments in gas-using equipment that they would come to regret when the price of gas increased — as increase it had to.) Since it is now clear that the price of gas had to rise sooner or later, the present British government prefers the rise to take place sooner, so that the discomfort caused by these price rises is not blamed on the increased privatisation that it is determined to sponsor. Readers interested in what the Thatcher government is trying to achieve can refer to J.D. Davis (1984, Chapter 5) for a very readable and thorough survey, the high point of which is as follows: 'One could certainly argue that in view of its past performance, the British Gas Corporation deserves better of the government'.

It is one of the theories of the BGC that there is not enough gas remaining in British fields to satisfy the demand that should exist towards the end of the present century. (The BGC's customer base is currently 16.5 million accounts, expanding by 300,000 per year). They therefore attempted to contract for additional supplies from Norway's Sleipner field, which holds 200 Gcm of proved reserves; but the present British government rejected the plan, claiming that the only thing that kept more gas from being discovered in and removed from British fields was the low price which gas producers were receiving.

The total cost of the gas that was to have been imported from Sleipner was reported to be 30 billion dollars, which could be expressed as a base price of 4.15 dollars/MBtu. This can be considered reasonable, and probably was acceptable to the Norwegians only because of the size of the deal. At the present time the average price of gas in the United Kingdom is about 3 dollars/MBtu, and this fairly low price has been roundly criticised by various free market advocates in the country. It has been claimed that the British Treasury desires higher gas prices because it will result in BGC paying higher taxes, while the Department of Energy wants higher prices in order to encourage the demand for nuclear-based electricity. Somewhere in all this there are some interesting economic problems which somebody in Her Majesty's Government believes that they have correctly solved. It has been suggested, for example, that the prospect of the absence of Sleipner gas in British pipelines will ensure that more gas will be found in the British North Sea, which in turn will mean more jobs for Britain, more tax revenue for the government and a sounder balance of payments — since imports of gas will not cause a drain on foreign exchange reserves. On the other hand, if the gas that is supposed to exist in the British North Sea does not exist, or is extremely high-cost gas, then a lot of people are going to be in trouble. Unfortunately, amongst those in dire straits we are unlikely to find the economists, politicians and civil servants who failed to understand that in these matters it is geology, and not economics, which has the last word.

France

France is a country in which the superbly educated energy bureaucracy fully comprehends both the economic and political significance of energy, and for the past 20 to 25 years have possessed genuine power in French governments. As things now stand all energy media are scheduled to decline in importance with the exception of nuclear power, and this includes gas. The French are, however, fairly flexible on this matter, and the likelihood is that if the excess supply of gas appears to be a long term thing, it will be taken advantage of — probably at the expense of coal. Table 8.3 lists the sources of French gas since 1960, and the 1990 predictions of the Josephe Report (Rapport du Groupe Long Terme sur l'Energie).

Today there is much speculation about the future intentions of Algeria and the Soviet Union. By 1990 the Soviet Union may be supplying as much as 35 per cent of French gas, and 4.5 per cent of its total supply of energy. Much of the tension associated with the Soviet role is manufactured in Washington DC, because in general the French consider the Soviets to be reliable, and if more Soviet gas

Table 8.3: Sources of French natural gas (all figures Gcm)

	1960	1970	1975	1981	1982	1990
France	2.7	6.3	6.7	6.5	6.2	3
Netherlands		2.6	8.2	8.1	4.9	4–8
Algeria		0.6	2.4	4.0	6.4	8.5
Norway				2.4	2.5	4
Soviet Union				3.9	3.4	10.5–12
Others				1.1	0.9	
Total	2.7	9.5	17.3	25.0	23.4	30–35.5

Source: Rapport du Groupe Long Terme sur l'Energie, Pétroles Information (Paris), 15 July 1983.

eventually becomes available it would not be surprising to see France increase the amount purchased from that country.

The situation with Algeria is different, and complicated by the fact that Algeria is now acutely aware that it has a limited supply of hyrocarbons, and from the point of view of achieving their development ambitions, these hydrocarbons must be depleted in an optimal manner. As a champion of the *prix juste* Algeria, in 1980, unilaterally raised the price at which it was exporting gas to France, which led to reduced deliveries and the French paying the 'old' price into blocked accounts. The dispute was finally resolved by the new French government of François Mitterrand, which resulted in the Algerians obtaining a higher price for their gas than most exporters, although as the gas price fell due to the manner in which it was indexed to oil, it drifted further and further from the original conception of a prix juste.

The problem for the French is that the concept of a prix juste still lies heavy in the air. Other clients of Algeria have run into similar trouble, and the basic question concerns what will happen when the oil market tightens again, and energy sellers start moving back into the drivers' seat. The Soviets also have their ideological quirks, but up to now they have tended to keep them separate from purely business transactions.

As with most major gas users, France is attempting to increase the amount of storage capacity to which it has access, relative to annual consumption. Storage capacity approximately doubled between 1978 and 1984, and it is scheduled to double again by 1990, when stocks may equal 42 per cent of annual anticipated demand. An important innovation in French crisis management is the system of interruptible contracts, which are used by many consumers with dual gas-oil or gas-coal firing arrangements: these consumers pay less for gas, on condition that gas deliveries may be suspended at short notice — presumably on emergency grounds. Plans are that by 1990 interruptible demand is to constitute 25 per cent of expected demand for gas.

What we expect is that by 1990 stocks and interruptible demand will be larger than the amount of gas obtained from both Soviet and Algerian sources, and so by drawing down stocks and interrupting deliveries to owners of dual-firing capacity, it would be possible to substitute for Soviet *and* Algerian gas for a year. However, since these countries have no reason to act jointly in natural gas politics, the conclusion must be that France is extremely well insured against a gas cut-off, and that in reality the French security problem — it it can in reality be called that — is minimal. As far as I am concerned, however, it is even less than that, because for Algeria (with its limited supplies of oil and its rapidly growing population) an interruption of gas sales to France, unless an alternative market was found in the United States, would be a catastrophe. Since a simultaneous cut-off of gas to both Italy and France is unlikely, Italy might be able to increase its purchases from Algeria, and transfer gas via pipeline to France. The same is true of other European countries.

West Germany

The consumption of gas in West Germany expanded at a rapid rate until 1979, but between 1979 and 1982 demand fell by approximately 17 per cent. In 1984 it recovered to 54.4 Gcm, but this was still lower than in 1979.

These figures conceal important sectoral differences. Consumption in the residential-commercial sector has never stopped growing, largely because of the increased demand for gas for space heating. A quarter of West German homes now use gas for space heating, and according to Czerniejewicz (1985) half of the new flats constructed in 1984 used a gas-fired space heating system.

The use of gas in electricity generation has decreased by half since 1979, and had it not been for the political opposition to nuclear energy, and the environmental problems caused by coal that are just coming to the attention of the German electorate, gas might have been almost pushed out of the electricity generation picture, except for some peak-load applications. As things stand, it is almost certain that the share of gas in this market will continue to decline from its present 16.5 per cent. At the end of 1984, 40.5 per cent of the gas used in West Germany was used in residential-commercial applications, 30 per cent in industrial applications and 13.5 per cent in other uses.

Because of environmental issues, and the inability of decision makers in West Germany to get a free hand to use as much nuclear power as they would like, there is a general tendency to discount or downgrade the security issues associated with Soviet gas. Germany has imported gas from the Soviet Union for ten years, and during this period the only interruptions in delivery were caused by cold weather, and these were later made up. Accordingly, it would be difficult to convince the energy bureaucracy and others responsible for future German energy supplies that they should lose the opportunity to sign advantageous new contracts if the Soviets could significantly increase the amount of gas reaching Central Europe.

Important amounts of gas now reach West Germany from Norway, where fields in the Ekofisk area, as well as at the Statfjord, Heimdal and Gullfaks fields supply 14 per cent of present demand. An important source of gas for West Germany will be the Troll field (considered below), on which Ruhrgas has concluded exploratory talks with Den Norske Stats Oljeselskap (Statoil); but now that the Sleipner deal with Britain has fallen through, and the Ruhrgas group had earlier expressed a desire to buy gas from that field, it is possible that Germany will get Sleipner gas. Germany has also been purchasing a small amount of gas from Denmark since 1984, and as production in the Danish North Sea increases, there should be an increased opportunity to purchase more at competitive prices.

The main supplier of gas to Germany over the past 20 years has been The Netherlands, which at present satisfies about 30 per cent of German demand. Holland has now decided that it has an extra 8.8 trillion cubic feet of gas to sell, which is more than the estimated amount in the Sleipner field, and it seems likely that Germany will end up with a part (and perhaps even a large part) of these supplies. Much of the export of Dutch natural gas is carried out by a firm called NAM-Gas Export, which acts for Shell-Esso and the Dutch organisation Gasunie, and since NAM-Gas also has equity in the German transmission companies Ruhrgas, Thyssengas and Deutsche Erdgastransport GmbH, German gas importers appear to be favourably situated relative to other potential buyers of Dutch gas.

ITALY, DENMARK AND THE NETHERLANDS

Italy is a country that at one time had very large quantities of domestic natural gas (located for the most part in the Po Valley). Denmark is a new gas nation. Both these countries expect to raise their consumption of gas in the near future, and perhaps by very large amounts. Italy's National Energy Plan of 1981 was designed to reduce the dependence on imported oil, and now with large amounts of Algerian and Soviet gas crossing the Italian border, i one of the few countries in the world to envisage the increased burning of gas in electricity generating plants. On the other hand Denmark intends gas to occupy only a transitory role in electricity generation.

Until recently assumptions have been that a steadily increasing amount of Algerian gas would find its way into the Italian pipeline system, but now — due to the peculiar history of the trans-Mediterranean pipeline — some doubts have been raised. Italy agreed to buy Algerian gas in 1977, and almost immediately after began construction of the 2,400 kilometer pipeline from the Hassi R'Mel gas field in Algeria to Bologna in Italy, where it would link up with other networks. The cost of the line was approximately 3 billion dollars (of which two-thirds was paid by Italy), and it was embarrassing for the Italians that the Algerian government refused to allow the line to be used for two years because of a dispute about price. Originally the Algerians wanted 6.11 dollars/MBtu at the Algerian border (i.e. 6.11 dollars/MBtu FOB), to which 1.20 dollars/MBtu would have to be added in the form of transit frees and royalties in Tunisia. Obviously, with Russian gas available in Central Europe at a price of 4.75 dollars/MBtu, the terms were unattractive; when Italy began to sign contracts for Soviet gas, the Algerians reconsidered their position. (An Algeria-Spain LNG deal was also interrupted, for four years, over a disagreement about prices.)

Originally the Italians intended to terminate gas deliveries from The Netherlands in the 1990s, as contracts expired, but now that the Dutch have decided to sell more rather than less gas, and have shown a pleasant flexibility where prices are concerned, it is likely that Italy will take more gas from that source. It is predicted that by the end of this decade 75 per cent of all Italians will be living in towns that have access to natural gas, and annual consumption may reach 40 Gcm/year.

Denmark also intends to break its dependence on oil as much as possible, and in theory this should be much simpler for Denmark than for Italy because of the presence of offshore gas resources. However, despite the fact that Denmark was one of the first countries to locate oil in the North Sea, gas exploitation has gone slowly. The main problem here is simple bureaucratic and political incompetence, which included the arguments of economists that it made more sense to sell Denmark's gas to the Continent than to finance a distribution system in Denmark that may cost more than 20 billion Danish crowns. The arguments were disregarded, and the primary Danish distribution network has either been completed or contracted for. The transmission pipelines in the Danish system run from the Jutland west coast near Esbjerg, and this line also supplies a north-south transmission pipeline running down the centre of the Jutland Peninsula. This line feeds into a number of underground caverns that are to form a main storage system, and also links up with the West German network. Distribution is accounted for by five regional companies, and during 1985 alone almost 1,000 kilometers of pipe was connected to electricity generating stations. By the end of 1985 at least 60,000 households had contracted to use gas, along with 75 industrial companies and at least 100 district heating companies. In June 1979 a law was passed which stipulated that the price of gas to homeowners could not exceed the price of oil (in terms of heating values), and until recently the price of gas to industrial users was two-thirds of that to households.

The state oil and gas company of Denmark, which is 100 per cent owned by the Energy Ministry, is conspicuously called Dong (Dansk Olie Og Naturgas), and has been a centre of controversy since its formation: on one occasion the government dismissed the entire board. Some people think that the company would like to establish and enjoy the prerogatives of Norway's Statoil, and this may well be so; but Denmark's hydrocarbon resources are small beer in comparison to those of Norway, and this is something that cannot be changed by any amount of the posturing and attention-courting that many Danish politicians are so fond of.

The Netherlands

The arrival of the energy crisis could, in theory, have made The Netherlands one of the most prosperous countries in Europe — if not the most prosperous. In 1973, The Netherlands was producing natural gas equivalent to about 1 million barrels/day of crude oil, of which about 45 per cent was being exported. Assuming that gas sold for only three-fourths of the price of oil, it can be claimed that The Netherlands was in possession of a bonanza. This bonanza was to a considerable extent squandered because it was spent on consumption and not investment, and its squandering has led to a much more dramatic question: what will happen to The Netherlands when its production begins to fall at a rapid rate, and nobody comes forward to claim that new reserves can be found if they are looked for with a little more skill and enthusiasm? Clearly, it is only a matter of time before the high standard of living to which the Dutch have become accustomed — but which is already falling for some groups in the community — must visibly decline for a large part of the population.

Certainly, without the constantly growing income from gas, public sector spending must eventually shrink drastically (since The Netherlands have already reached a point where it cannot continue to borrow at the present pace), and when public sector spending falls most of the people in the country will be negatively affected.

Few countries in the world use as much gas as The Netherlands, and this was brought about by keeping the price of gas well below that of oil. It was many years after the initial introduction of gas into the Dutch economy before a determined attempt was made to economise on the use of gas. Low-priced energy was as much a part of the comprehensive social welfare system as the public broadcasting of soft-drug prices that took place on the Amsterdam radio. More than 80 per cent of private households use gas for heating, and about 80 per cent of industry uses gas. According to some observers, gas was mistakenly considered by The Netherlands to be virtually inexhausible, but this is not a correct judgement. The assumption that was made in the early 1960s was that while gas was exhaustible, it would last a long time, and when it was gone, or going, the superb Dutch technicians and administrators would find something just as satisfactory. Unfortunately, if there is another form of energy standing by that can provide future residents with the kind of benefits that gas has provided Holland, nobody is saying anything about it. Instead, plans to use more coal are constantly updated and revised. Some of the best coal in the world is located offshore of The Netherlands, although these resources are probably limited in quantity.

It has already been mentioned that The Netherlands is potentially of great importance to Western European energy security because of the superb storage facilities offered by the depleted portion of its gas fields; but in addition the Groningen gas province can still develop considerable surge capacity — that is, the flow of gas from Groningen fields can be increased very rapidly. This is not true, for example, of Norwegian fields, largely because of their location and the complexity of the pipeline systems associated with these fields. This being the case, it may be optimal to deplete the gas fields of the Netherlands at a slower pace than the Dutch desire, and if the real world functions in the same manner as the textbook world, The Netherlands gas would be entitled to a premium on these grounds.

NORWAY

Much of the above discussion has been about Norway, because it is impossible to discuss the energy situation in Europe without constantly bringing Norway into the picture.

Considered in relation to its population, Norway may be the most energy-rich country in the industrial world; and objectively speaking, since the early 1970s, Norway has been in an economic and social position that is more favourable than that of any other industrial country. This does not mean that the Norwegians have succeeded in building a welfare state paradise commensurate with their huge natural resource endowments, but on the average both this and the next generation of Norwegians should be able to look forward to lives that are largely free of the nagging insecurity that many other Europeans feel at the present time.

In 1985 the Norwegian government was able to collect about 46 billion Norwegian crowns in petroleum (oil and gas) related taxes. At the present exchange rate this amounts to 6 billion dollars which, while it may not sound like very much in terms of the tax revenues of most European countries, is a great deal for a highly developed country with a fairly small population that is already rich. It has been said that the seven fat years that began in 1979 are due to be followed by seven lean years, beginning now, but they will be lean only in comparison to what they might have been had the price of oil remained high.

This year the petroleum-related income of the Norwegian government could fall to below 30 billion crowns. The reason is the collapse in the oil price, which to a certain extent has been made more extreme by the oil production policies of Norway and the United Kingdom. These countries have reacted to declining world demand for oil by raising their production, despite the fact that the most favourably endowed of the OPEC countries (Saudi Arabia, Kuwait and the United Arab Emirates) have long possessed the capacity to undercut the price at which other exporters offer to supply oil, and to take virtually any share they want of the oil market. This observation especially applies to Norway, whose oil is among the most expensive in the world to produce. From 1983 to 1985 Norwegian oil production increased from 30 Mt/y to 40 Mt/y (= 0.8 Mbbl/d on average), and intention was for it to reach from 60 Mt/y to 80 Mt/y by 1990. This means that Norway would have appropriated a sizeable part of the expected world increase in oil consumption over that period — which hardly would increase its popularity with those OPEC countries that are desperately in need of cash.

Known Norwegian oil reserves permit at least 25 years of extraction at present rates, while gas reserves allow at least 100 years of production at the present annual output. Proved gas reserves increased by nearly a fifth over the period 1975–85, and in 1985 gas output was approximately 24 Gcm, which in calorific content came to 0.43 Mbbl/d of oil. All of Norway's gas output is exported, as compared to 85 per cent of its output of oil.

Geographically, mainland Norway covers about 324,000 square kilometers, and has a population of 4.5 million people. The birth rate is low, and since Norway — together with Finland and Iceland — are the countries in Western Europe that are most unfriendly to immigration, it seems unlikely that it will ever rise to much over five million. According to existing conventions, Norway has the right to lay claim to about 2,000,000 square kilometers of sea and seabed off the coasts of Norway, and it is in these regions that the oil and gas have been found. Thus far only about 15 per cent of the Norwegian continental shelf has been explored, mostly in the Norwegian part of the North Sea, but it is clear that in the Arctic Ocean (off the coast of northern Norway) rich hydrocarbon resources are likely to be found. These resources will be the most expensive of their kind in the world, however, and the installations that will have to be constructed to exploit them belong to the very frontier of petroleum technology.

The gross national product (GNP) of Norway increased by three per cent in real terms in 1983, four per cent in 1984, and 3.5 per cent in 1985. By present European standards these are high rates of growth, but they are still under the increases that were expected only a few years ago. Wage costs have been rising very rapidly, and in general private consumption has grown at a faster rate than GNP. Investment has also been increasing fairly rapidly, especially investment associated with the oil sector. The unemployment rate in Norway is the second lowest in Europe (after Switzerland), and comprehensive social welfare schemes virtually guarantee that unemployment does not entail a large fall in income. Norway went heavily into debt during the 1970s, but the second oil price rise lowered the profile of this problem, since it graduated Norway from the ranks of the rich to those of the very rich. A conscious effort has been made in Norway to raise the competitiveness of domestic industry, and thus avoid (pseudo) dilemmas of the type known as the Dutch Disease (in which expansion of a natural resource sector leads to serious problems in other sectors due to increased wage costs and an appreciating exchange rate). The Norwegian government also, at one time, possessed ambitious plans for channelling a large part of the Norwegian oil revenues into investment abroad, but fortunately — for Norway — abandoned this ill-conceived scheme before it caught the imagination of those academic and political cranks who might have turned it into an economic disaster.

The Gold Block and Troll

For the last few years the Norwegian exploration strategy has concentrated on a search for new oil fields, rather than new gas fields, but it has suddenly become more difficult to find new oil. The opinion now is that in the Gold Block — or 34/8 as it is formally designated — where at one time it was believed that oil worth 500 billion Norwegian crowns was located, there is hardly any oil at all: thus far the only thing that the Norwegians have to show for millions of dollars of exploration expenditure is a huge dry hole, although some gas and condensates have been located.

With the results more or less established where 34/8 is concerned, there is a great deal of talk now in the petroleum bureaucracy about pushing forward the development of the promising oil discovery known as Snorre, that was made in blocks 34/4 and 34/7. Snorre is said to contain 100 million tons of oil, and can be in production before 1994. There is also some talk about an earlier exploitation of the Brage field, as well as the exploitation of some marginal fields. One of the problems with developing Brage, however, is that taxes on this field might have to be lowered, because at present tax rates and oil prices, operations in these fields are definitely unprofitable — especially when set-up costs are considered. Regardless of which course of action is taken, however, 'oil-Norway' is on its way to becoming 'gas-Norway'. The percentage of total petroleum reserves consisting of gas is increasing as the percentage of new natural gas reserves increases at an even higher rate. It has been suggested that there is actually much less offshore oil in Norway than originally believed, and if this is so it might make sense to extend the life of reserves by sharply decreasing the rate of production of oil — particularly now that oil prices are low.

The failure of the Gold Block came at an exceptionally bad time, because after the failure of the Sleipner gas deal, the Norwegian government had hoped to show potential customers that there was really no hurry in developing the northernmost gas fields. But now there is a hurry, because having accustomed themselves to a high standard of living, the Norwegians have no desire to reverse this process, nor would it be wise for a political party to espouse such a line. The present Norwegian government has pledged itself to push the development of the huge Troll field, and as pointed out below, a deal that has been called the largest in Norwegian history appears to have been arranged with a consortium of West European buyers. (Note the 'appears'. These transactions are never certain until the gas is actually flowing.) The once important subject of a large transaction with British Gas is not discussed much in public these days, but there are still a number of influential people in natural gas who assume that Britain will have sizeable excess demand for gas at the beginning of the 1990s, and this excess demand will grow progressively larger. Also, after the next general election, Britain may have a government with a different point of view on the geological characteristics and potential of the British North Sea, in which case a deal between Norway and Britain might be possible. Here it should be remembered that Norway has the gas to supply Britain, even if it does arrange other huge transactions in the meantime, since the reserves of the Troll field alone are 1.2–1.3 Tcm, at least.

The contract with the West European Consortium will be worth about 68 billion dollars, and calls for the delivery of 450 Gcm of gas over 30 years from both the Sleipner and Troll fields. The start is to be in 1993, and deliveries are to be stepped up throughout the 1990s to provide 20 Gcm annually between the years 2000 and 2020. The participants in the transactions are Statoil on the Norwegian side; and on the Continent Ruhrgas, Thyssengas and BEB of West Germany; the Dutch state company Gasunie; Distrigaz of Belgium; and Gaz de France. Statoil owns only about half of these fields: Exxon is a major owner of Sleipner, and Shell has the same position with Troll. Thus the so-called victory for Norway and the Reagan Administration (which has never ceased to push for more Norwegian — and less Soviet — gas in the Western European energy picture), is also a victory for several very large oil companies, among the most prominent of which are Exxon and Shell. It might also be termed a victory, however, for steel manufacturers and shipyards in Northern Europe which will benefit from orders from this project. There are also people who say that this is the definitive breakthrough for natural gas in the Western European energy economy.

Norwegian sources once indicated that about 6 billion dollars in capital investment would be needed to develop the Sleipner field.

Now, with both Sleipner and Troll to be developed, the problems associated with the financing of this investment seem to have disappeared. Among other things, there is no longer any doubt about the economic feasibility of a pipeline from the north of Norway to Zeebrugge.

Where Troll is concerned, the problem is with the technology required to exploit this and similar fields efficiently. The hazards of offshore work in the North Sea are well known throughout the oil world, and as the movement is made from the North Sea towards the Arctic Ocean, they are not expected to diminish, nor are investment costs. Early offshore platforms, such as the Ekofisk complex, were in only 24 meters of water, but even so had to be prepared to resist winter storms where wave height could reach 30 meters. When production platforms reached 150 meters, production and living quarters were integrated. The most northerly platform in the North Sea, which is in BP's Magnus field, is located in 185 meters of water; the world's tallest and largest platform is in the Gulf of Mexico, and sits in 312 meters of water: this distance is the same as the height of the Empire State Building. But in the Troll field the average water depth is 350 meters, and conventional offshore platforms — if they can be constructed — would need 800,000 tonnes of concrete, making them the heaviest mobile objects ever built. It is estimated that a platform of this description would currently cost at least seven billion dollars; and one of the dilemmas that might have to be faced in the Troll field is that since the field is spread out over a large area, a number of platforms might be needed.

It has been suggested that one of the ways to approach this problem is to abandon previous technology, with its fixed structures, and to build platforms designed to move with the waves; or to locate most of the production installation on the seabed, and operate it by remote control. This latter type of technology is being tested in the North East Frigg field, which is a marginal field where it was not considered profitable to build a conventional fixed platform. The heart of the system is a small unmanned platform on a compliant column, fixed to the sea bed with a universal joint, and featuring six sub-sea wellheads to monitor and control the output of gas.

With all this speculation about technology, it is clear that as yet nobody knows how much it will cost to exploit the Troll field. And this opens two possibilities. One is that Troll will not be exploited until very late, and as a result there will be substantial excess demand for gas in Western Europe in ten years that will drive the price of gas sky high; or that the buyers of gas will ascertain that things are not developing the way they hoped in Norway, and rush to contract for supplies from the Soviet Union, Algeria and perhaps the gas exporters of the Middle East. In all likelihood it was for these reasons that the Norwegian government, which once insisted on a deal for Sleipner before it would consider developing Troll, acted now so as to have a deal on Troll gas concluded as fast as possible. Preferably before the price of oil goes even lower, and thus weakens Norway's bargaining position; and at a more abstract level, before the end of the war between Iran and Iraq, when the gas producers of the world might have to face the unappetising prospect of competing with the potentially huge exporting capacity of Iran.

APPENDIX: ALGEBRAIC COMMENT ON SECURITY OF SUPPLY IN THE WESTERN EUROPEAN GAS MARKET

Much of the discussion about the security of supply of natural gas in Europe is not about security but about selling gas — Norwegian gas, to be exact, whose owners are Norwegians (and perhaps Americans). I can understand the behaviour of the Norwegians and their business partners in this matter; but I cannot understand why many of the potential clients remain passive while the richest country in Western Europe merrily insists that it has a legitimate right to get even richer.

In the not too distant past, Norway wanted to put a floor price of 6–7 dollars/MBtu on its gas, or more, while at the same time Soviet gas was selling for about 4 dollars/MBtu. If readers wonder what I am getting at, they can examine a recent paper by J.P. Stern (1986) in which he uses the expression, '. . . and the choice may be between Soviet gas and high-cost Norwegian supplies'. As pointed out in my IAEE lecture at Cambridge University in 1982, I do not feel that there are any extensive security risks with Soviet gas; but that is strictly an off-hand opinion about an issue in which I have very little interest: whether Soviet gas is used to a greater or lesser extent is something that will have to be determined by the politicians of Western Europe, not by academic economists. What I am interested in, however, is the type of article published by Alan Manne, Kjell Roland, and Gunter Stephan (1986), where with the aid of some graphics from one of the more irrelevant branches of academic economics, the Pure Theory of International Trade, these scholars suggest that they are in possession of a valid technique for judging security issues.

In cost-benefit analysis the crucial dimension is time. This is not considered in a meaningful way by Manne, Roland, and Stephan — or MRS, as I will call them. To understand this issue, suppose that it is not just likely but certain that there will be a disruption in the supply of external gas to Western Europe in 20 years. Still, there could be a sound economic argument for purchasing this gas today if it were much less expensive than Norwegian gas, because the difference between the total cost of Norwegian gas and 'other' gas, accumulated over 20 years, could subsidise the construction of an alternative energy system (gas or otherwise), and still permit a windfall for consumers of gas.

A short example will be constructed below based on this idea, but before doing so let me say something about the 'probability of disruption' used in the MRS paper. It is a subjective probability. In the example below I am going to assume that for some years the disruption probability is unity, but that does not mean that a disruption must take place: it means that it is strongly believed that a disruption will take place in a certain year, and if it does not take place that year, it is equally strongly felt that it will take place the following year, and so on.

I shall use the net benefit measure for linear supply and demand curves that was employed by MRS. Though it makes no difference what kind of supply and demand curves we have, because most people outside the Stanford Institute for Mathematical Studies in the Social Sciences would never consider drawing supply and demand curves for natural gas of the type in the MRS paper. These curves are strictly irrelevant (and equally irrelevant is tariff and quota theory when applied to those countries that do not have significant indigenous supplies of gas to protect — which is certainly true of most of the potential gas importers of Western Europe). From my point of view the key issue is the inter-temporal assignment of benefits and costs resulting from the purchase of external gas, and the evaluation of their present value. I will take the net benefit (NB) function as $\frac{1}{2} [(1-p)a - p(b-a)]x^2$, where p is probability, and for the constants a and b, $a,b > 0$, and $b > a$. Next I make the assumption that for a given period in the immediate future, t_1, the subjective probability of a supply disruption is zero; and for all time after that it is unity. We thus have:

$$NB = \frac{1}{2} ax^2 \qquad 0 \le t \le t_1 \qquad (p = 0)$$

$$NB = -\frac{1}{2}(b-a)x^2 \qquad t_1 < t \le \infty \qquad (p = 1)$$

If it can be assumed that the units on the horizontal axes of the diagrams in MRS which say gas really mean gas/year, then the expression that interests us is:

$$NB = \int_0^{t_1} \frac{1}{2} ax^2 e^{-rt} dt + \int_{t_1}^{\infty} -\frac{1}{2}(b-a)x^2 e^{-rt} dt$$

This yields:

$$NB = \frac{1}{2} ax^2 \left[\frac{1-e^{-rt_1}}{r}\right] - \frac{1}{2}\left[\frac{(b-a)x^2}{r}\right] e^{-rt_1}$$

The second expression on the right hand side of the equation immediately above (including the sign) is the present value at time t_1 of a dis-benefit in the form of a perpetuity, discounted back to time $t = 0$. The right hand side of this equation can now be simplified to:

$$NB = \frac{x^2}{2r} [a - be^{-rt_1}]$$

Discounted net benefits are positive when:

$$a - be^{-rt_1} > 0$$

Or, after simplifying:

$$\frac{a}{b} > \frac{1}{e^{rt_1}}$$

Remembering that $b > a$, there is clearly a value of t_1 that satisfies this inequality, which means that even if we believe it certain that there will be a dis-benefit in all future years after t_1 because of a supply disruption, they are offset by the benefits gained between now and the year t_1.

The example was carried out with the aid of MRS' benefit function which — of course — is completely arbitrary; but even so

I think it serves to illustrate the point I am trying to make in this note. In the real world, money values would be used instead of a benefit function; and it was because money values were used that many countries in Western Europe chose Soviet gas instead of Norwegian gas.

9

Economic Theory and Natural Gas Pipelines

The purpose of this chapter is to provide the reader with an introduction to one of the most important, but neglected, subjects in energy economics: natural gas pipelines. Many of the concepts also apply to pipelines carrying crude oil or oil products. For pedagogical reasons it is necessary to include a considerable amount of mathematical material in this chapter, but the general reader will find that the discussion presented in the first two sections is only slightly technical, and covers both pipelines and depletion of gas fields. More general surveys of the natural gas markets can be found in an article by David Hawdon (1985), in which pipelines are referred to in an appendix, in A.R. Khan (1984), K. Roland (1984) and O. Fugelberg (1984).

In the third section of this chapter I begin an analysis of pipeline economies that is based on a brilliant article by H.B. Chenery (1949). A short summary of this paper can be found in V. Smith (1961). Neither Chenery nor Smith paid sufficient attention to the effects of volume when considering production from natural gas reservoirs, but this hardly detracts from the importance of their work: Smith is the only economist with the correct approach to this problem. I have accordingly made a special effort to illuminate the intertemporal aspects of exploiting natural gas reservoirs. As the reader will soon notice, this means that we must stress certain issues in production theory that most microeconomic textbooks make a point of disregarding. But see also Alchian (1959).

DEPLETION OF GAS FIELDS

The first problem concerns the depreciation of a non-associated natural gas reservoir (i.e. a reservoir that does not contain appreciable quantities of oil) or an entire gas field. If we assume that the presence of a natural gas reservoir has been confirmed, investments are initiated in such things as drilling equipment and production platforms (if the deposit is offshore). Once a certain amount of initial investment has been made, the gas will start flowing, gradually building up to a peak or plateau. During the period of plateau production and afterwards, the costs associated with depleting the reservoir will be mostly labour costs and costs related to the depreciation of equipment. If it is felt that the technical and economic conditions justify prolonging plateau output, investment may take place in new wells or compression units to substitute for falling natural pressures in the reservoir. In Figure 9.1 this contingency is shown by dashed lines.

Figure 9.1: Investment and production profile of a natural gas field

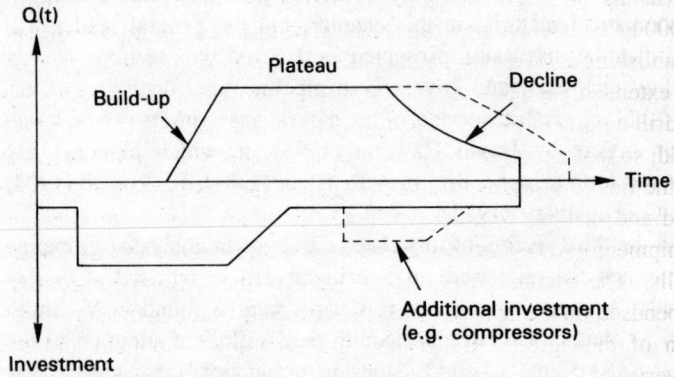

A gas field is sometimes described as having a 15-year depletion period if its annual output on the plateau is one-fifteenth of total recoverable reserves. This plateau output level might be held for considerably less than 15 years (e.g. ten years), although gas might flow for more than 15 years, taking into account both the build-up and tail-off (i.e. decline) periods. The advantage in looking at the problem in this manner is that it allows attention to be focussed on two important variables — the height of the plateau (i.e. plateau output); and its length. The same is true for oil.

The amount of gas that can be considered recoverable is that which can be directed from a reservoir into surface pipelines. Most of this movement is made possible by a pressure difference between the gas reservoir (or trap, as it is sometimes called) and the pipeline entrance that results from reservoir pressures of up to 10,000 pounds per square inch (= 10,000 psi). A gas reservoir can often be emptied of more than 80 per cent of its contents, as compared to an average of 32 per cent of oil, with as much as 50 per cent of this attributable to gas expansion alone. For instance, Algerian gas comes out of its wells at such high pressures that, as yet, no compressor stations are needed to move the gas the more than 400 miles to the Tunisian border. As for the pipeline, delivery can take place as long as a suitable pressure differential is maintained between the pipeline entrance and the point of delivery. In general pipeline networks consist of high pressure transmission pipelines and low pressure distribution systems, with the latter buying from the former. The discussion in this chapter concerns almost exclusively transmission pipelines.

If we consider on-shore gas fields (where a field can be taken as consisting of a number of reservoirs, while all the fields within a 1,000 to 10,000 square mile area comprise a basin), the first step in establishing a system which can delivery large amounts of gas over an extended period — according to the terms of the contract — is to drill a number of exploratory wells over the estimated area of the field, so that an estimate can be obtained of the recoverable reserves of the field. With this estimate available, the gas in the field can be sold and drilling completed. Should it be necessary, compression equipment can be installed to boost pressures in the vicinity of the wells. The number of wells drilled and the drilling programme depends upon the desired production profile, which in turn is a function of the agreement between the owners of the field and the buyer(s) of the gas, and almost always involves a uniform or increasing rate of delivery over a comparatively long period.

A constant level of production or sales is a datum, and not always a fact. In 1972, in the United States, declining production from existing gas fields led to industry temporarily laying off 1.2 million workers so that gas could be diverted to top priority residential users; but even so many communities were on the fringe of 'gas-outs', that would have left tens of thousands of homes cold. Conversely, recent practice has often been for natural gas consumers to refuse gas that was contracted for, despite take-or-pay contracts. In a situation charaterised by the gas bubble (i.e. a temporary over-supply of gas), the courts have tended to favour the purchasers of gas.

Of particular importance in any discussion of natural gas in the United States is the declining productivity of new gas wells, which means a larger number of wells and sharply increasing expenditure to obtain equal increments of output from the exploitation of existing or new gas fields. This is not just due to such things as well-clogging, falling inter-well pressure differentials, and the increasing density of wells in a particular gas field (which means that with a fixed amount of gas in a gas field, new wells will tend to feature a decrease in output), but the actual scarcity of gas down to the maximum depth at which gas is currently found in most of the known gas basins.

In the United States the number of wells drilled has increased from 9,850 in 1977 to more than 16,000 in 1982, and according to A.R. Khan (1984) average well productivity has declined from 115 million cubic feet per year (= 115 Mcf/y) to 65 Mcf/y. This situation directly implies that, on the average, maintaining a constant level of production will involve finding and bringing into production an increasing number of wells. There are now at least 230,000 producing wells in the United States, and if the current rate of production decline continues, then at least 450,000 wells will be required by 1990 to provide the 15 trillion cubic feet per year (= 15 Tcf/y) of non-associated gas that is now being supplied. Figure 9.2 indicates average gas well production in the United States over the period 1965–84.

Figure 9.2: Average gas well production in the United States (1965–84) (in millions of cubic meters — Mcm)

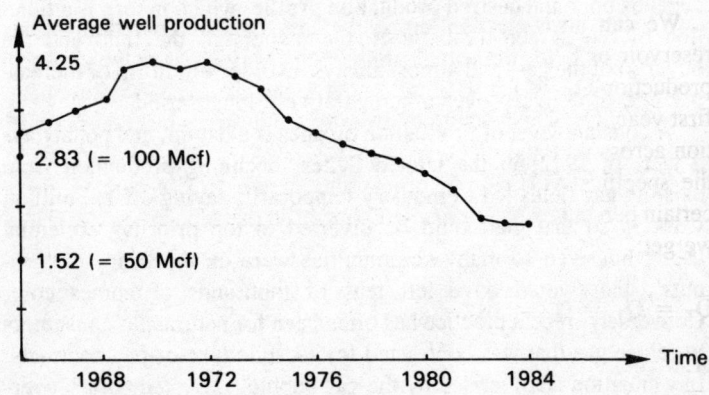

Note: Preliminary estimate for 1984.
Sources: *Oil and Gas Journal* (various issues)

In the light of the above discussion, I would like to derive in a less complicated, and more pedagogical, manner a relationship that, as far as I know, was introduced into the economics literature by P.W. MacAvoy (1962). What it involves is the decline in pressure in a reservoir or gas field, and the influence of this decline on the production capacity that must be provided to maintain a uniform flow of gas. In MacAvoy's analysis the decline in maximum capacity, per period, is taken as equal to the amount of gas, in percentage terms, removed from the deposit of gas in the first period. For instance, if there are 100 units of gas in the deposit from the beginning, and five per cent — or five units — is removed in the first period, then the maximum capacity in the second period is 95 per cent of that in the first, and so on. I prefer to see this constant depletion rate introduced as a special case, just as I prefer to avoid MacAvoy's practice of working with fractions. Thus, taking a_i as the (percentage) amount by which maximum capacity declines in period i, and accordingly $\overline{Q}_i$ as capacity output, we get:

$$\overline{Q}_2 = \overline{Q}_1 (1 - a_1)$$
$$\overline{Q}_3 = \overline{Q}_2 (1 - a_2) = \overline{Q}_1 (1 - a_1)^2$$
$$\text{If } a_1 = a_2$$

$$\vdots \qquad \vdots \qquad (1)$$

$$\overline{Q}_T = \overline{Q}_{T-1} (1 - a_{T-1}) = \overline{Q}_1 (1 - a_1)^{T-1}$$
$$\text{If } a_1 = a_2 = \ldots -a_{T-1}$$

We can now specify that in the last year of operation of the reservoir or field, the actual production Q_T is equal to the capacity production $\overline{Q}_T$, which in turn is equal to the actual production of the first year, Q_1. This is done in order to get a uniform actual production across periods, or $Q_1 = Q_2 = \ldots = Q_T$. Next we introduce the specification that the production in the first period Q_1 is a certain percentage x of the initial content of the deposit, or xM. Thus we get:

$$\overline{Q}_T = Q_T = Q_1 = xM = \overline{Q}_1 (1 - a_1)^{T-1}$$

or: $$\frac{\overline{Q}_1}{M} = \frac{x}{(1 - a_1)^{T-1}} \qquad (2)$$

In MacAvoy's analysis $a_1 = x$. Thus, if a_1 = five per cent, and T is 20 years — or five per cent of the initial capacity of the reservoir is to be supplied for 20 years — then we would have:

$$\frac{\overline{Q}_1}{M} = \frac{x}{(1-x)^{T-1}} = \frac{0.05}{(0.95)^{19}} = 0.1325 = 13.25 \text{ per cent}$$

In other words, initial well capacity should be sufficient to produce 13.25 per cent of the deposit. If this is so then, due to the decline in pressure, that capacity will be able to produce five per cent of the initial capacity of the deposit in the final period. There are clearly other ways to handle this problem. One is to add wells or compression equipment later to compensate for capacity declines. In point of truth, however, I do not know how reservoir engineers choose between various options. Note also that we do not have to have $a_1 = a_2 = \ldots = a_T$. They can be different, and in some cases related through a decline curve.

NATURAL GAS PIPELINES: INTRODUCTION

This section presents a simple introduction to natural gas pipelines, and at the same time an attempt will be made to touch on some issues that will receive a technical exposition in the next two sections. We have already noted that as gas flows out of a deposit, the pressure in the deposit decreases; and if the pressure falls in such a way that the gas cannot be captured and routed to its purchasers, then equipment called compressors must be installed to boost the pressure.

Naturally, if we are thinking in terms of transporting the gas over long distances, compression equipment is going to be necessary regardless of the pressures in the deposit. Also, as alluded to in the previous section, compression equipment may not be initially necessary for the short distance transmission of natural gas, but as pressure falls in the deposit, or throughput (i.e. output) is increased, compressors must be introduced.

Figure 9.3 shows the situation for a typical initial portion of a transmission line. The pressure at the gas source is p'_0, and falls to p_0 at the entrance to the first compressor. It is raised to p_1 in the compressor, and falls to p_2 at the entrance to the second compressor due to friction in the pipeline, etc. Compressor stations need not be the same size, and inlet and outlet pressures at each of the stations need not be the same; but there are often important economies to be

Figure 9.3: Gas source, compressor stations, pipeline stretches and pressures in a simple natural gas transmission pipeline

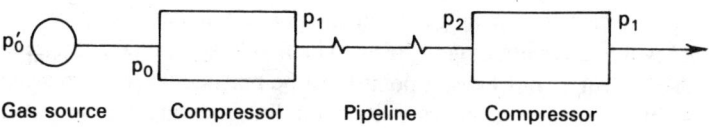

gained from uniformity in these matters.

Good examples of existing and proposed pipeline systems are the pipeline from Australia's North West Shelf to Perth — Western Australia; and a proposed line from Norway to the southern tip of Sweden. This line could supply Swedish buyers, and/or purchasers in Denmark and Germany by extending the pipeline to these countries. The Australian pipeline is 26 inches in diameter and runs 940 miles (= 1,504 kilometers). It features five compressor stations, and 50 mainline valves, with both stations and valves controlled from Perth by a computer that can monitor 1,000 readings/second. There is one compressor station about every 200 miles, although as far as I can tell, the average spacing of compressor stations, taking pipelines in all parts of the world into consideration, is about 100 miles.

In the case of Sweden, the proposed line would be over 1,700 kilometers in Sweden alone, and the approximate cost would be four billion dollars (1984 dollars). The pipe would be 50 inches (= 1.5 meter) in diameter, and the thickness of the pipe 1.6 centimeters. Each length of gas pipe would be 12 meters, and two of these lengths would be welded into 17 ton packages before being put into the trench. The largest pipeline diameter that is considered feasible at the present time is 64 inches, although a few years ago some observers were predicting that 100-inch diameter pipelines would be under construction before the end of this decade. Some portions of the Soviet (Yamburg)-Western Europe pipeline system now employ 56-inch pipe, and in some of the proposed pipelines between Alaska and the 'lower forty-eight' — via Canada — 56-inch pipe with a wall thickness of 0.54 inches and a line pressure of 1,080 pounds per square inch (psi) was considered appropriate for at least one of the stretches. At the present time though the arrangements that seem to be favoured are 48-inch lines with a wall thickness of 0.54 inches and a line pressure of 1,260 psi, or 48-inch lines with a wall thickness of 0.72 inches and a line pressure of 1,680 psi.

Figure 9.1 indicates that there might be a trade-off between line pressure and pipe diameter. In most economic literature line diameter and pipeline pressure are taken as the two principal factors of production in moving gas from one place to another in a pipeline, and within a certain range appear to be highly substitutable for each other. Clearly, however, a point must be reached where increasing the line diameter cannot substitute for line pressure (or, to put it another way, compressor capacity), and this point appears sooner rather than later as the length of line is increased. Similarly, compressor capacity cannot substitute indefinitely for line diameter as throughput is increased. One of the weaker aspects of the existing economics literature treating this topic is that it does not place enough emphasis on the lack of substitutability in certain situations.

It is well known, however, that increasing the diameter of a pipeline will permit important economies of scale if sizeable quantities of gas are transported. The carrying capacity of a gas pipeline is proportional to its interior cross sectional area, which is πr^2, where r is the radius of the pipe. The amount of pipeline material (e.g. galvanised steel) used is proportional to the circumference of the pipe, which is given by $2\pi r$. If the input of steel is halved, the carrying capacity of the pipe falls to less than half the original amount; while if the input of steel is doubled, the carrying capacity is more than doubled. Obviously, there are limits to the scale economies that can be achieved by increasing the diameter of the pipe, and these limits are determined by — among other things — pressure differentials, length of the pipeline, etc.

What will be done now is to try to systematise, at an elementary level, some things that will be taken up in the next section in more technical discussion. To begin with we have:

$$Q = Q(p_1, p_2, D, L) \qquad (3)$$

Output (at the compressor entrance) is a function of inlet pressure, outlet pressure, pipe diameter and length of the pipeline. The diameter used here is the inside diameter. The horsepower requirements of the compressors are:

$$H = H(p_1, p_2, Q, L) \qquad (4)$$

An expression that is often employed when discussing required compressor capacity is p_1/p_2, which is the compression ratio. Equation (4) is thus occasionally written $H = H(p_1/p_2, Q, L) = H(R, Q, L)$, where R is p_1/p_2.

As most economists appreciate, two equations will not give the optimal values of five variables: $p_1, p_2, D, L,$ and H — assuming that Q is given. Instead, trial and error methods — together with these two equations — sometimes make it possible to get closer to an optimal solution. This is because of the power of modern computing equipment, which in some situations permit sheer calculation to substitute for the absence of a complete model or algorithm. But as the reader will soon see, I prefer heroic simplification to calculation.

In considering the costs of a pipeline, it is useful to think in terms of the cost of pipe and of laying the pipe; the cost of the compressor stations (which is the cost of constructing the stations and installing all the equipment they contain); the fuel cost of running the compressors, where the fuel is usually gas from the pipeline (although in the case of the proposed line for Sweden, electricity would be used to drive the compressors); and the maintenance cost of the line (pipes, valves and compressor stations) after the line is constructed, along with certain supervisory and engineering costs. Putting this kind of information into equation form is not easy, and in the one case where I have seen it attempted it was a resounding failure; but even so it might be useful to organise the analysis around some simple concepts. For example, the weight of a foot or meter of the pipe is a function of diameter and thickness. The cost of laying the pipe, per mile or kilometer, is usually reckoned as a function of weight and the kind of territory that is being traversed by the pipeline, assuming that all else is given and constant. Compression costs are almost certainly a function of the horsepower of compressors but the use of accessories such as coolers, chillers, or heaters will raise the cost. There is a great deal of information compiled on the cost of maintaining pipelines and compressor stations, and most of this information can be easily interpreted. An item of some relevance here is that once we know the investment costs associated with a project, it becomes a simple matter to turn this lump sum cost into a series of annuity-like annual payments that can be called a capital cost. (I have gone into this matter in some detail in my book *The Political Economy of Coal*, but it is also referred to below.)

We have now reached a point where another mathematical exercise must be carried out. As noted earlier, I am particularly concerned about bringing the volume of gas into the analysis in a meaningful fashion, since natural gas is usually traded in terms of a certain amount annually over a given time horizon T. Operationally this means, among other things, that a flow of costs and benefits from the project must be evaluated at one point in time — usually when construction of the project commences. (For the most part, in this chapter, I am concerned with costs.) This evaluation generally turns on the concept of present value, and in order to obtain a result that will be needed later I will postulate a stream of payments, each amounting to A, being made at the end of every year over the time horizon T. With a discount rate of r (which might be the interest rate), present value C is:

$$C = \sum_{i=1}^{T} \frac{A_i}{(1+r)^i} = \frac{A_1}{(1+r)} + \frac{A_2}{(1+r)^2} + \ldots + \frac{A_T}{(1+r)^T} \quad (5)$$

I will set $A_1 = A_2 = \ldots = A$, and approximate this expression by:

$$C = Ae^{-r} + Ae^{-2r} + \ldots + Ae^{-Tr}$$
$$= Ae^{-r}(1 + e^{-r} + \ldots + e^{-(T-1)r}) \quad (6)$$

Be careful to note here that the first payment takes place at the end of the first period. If both sides of (6) are multiplied by e^{-r}, and (6) is subtracted from the expression that is obtained, and simplified, we get:

$$C = \frac{A(1-e^{-rT})e^{-r}}{(1-e^{-r})} = \frac{A(1-e^{-rT})}{e^r(1-e^{-r})} = \frac{A(1-e^{-rT})}{e^r-1}$$
$$\approx \frac{A(1-e^{-rT})}{r} \quad (7)$$

$e^r \approx (1+r)$ when r is small, and with the fairly realistic values of $T = 20$ and $r = 0.06$, $rT = 1.20$. This is not small enough to make this approximation useful. Moreover, as the reader can also verify by straightforward integration:

$$\frac{A(1-e^{-rT})}{r} = \int_0^T Ae^{-rt}dt \ (= C) \quad (8)$$

The integral is the continuous version of (5) and (6), which is hardly surprising; but what is surprising to some people is that the lower limit of the integral is zero instead of unity, since in (6) the first payment took place at the end of the first period. Thus, if there is a payment at the beginning of the first period, it must be added to (8). I will employ this observation later in this chapter, under

Average and Marginal Cost, when I examine the effect of volume on costs in the context of a simple extension of neo-classical production theory. The present discussion can be concluded by presenting the discrete approximation for the left hand side of (8), where for simplicity C is put equal to unity.

$$A = \frac{r}{1-e^{-rT}} \approx \frac{r(1+r)^T}{(1+r)^T-1} \tag{9}$$

Here we have the familiar expression for an annuity, which is roughly defined as the annual payment over a period T equivalent to a single amount C at one point in time, usually the beginning of the period in which the first annual payment is made. With $C \neq 1$, the two right hand expressions in (9) must be multiplied by C in order to get the value of A.

PIPELINE ECONOMICS: A NEO-CLASSICAL FORMULATION

The next project will be to examine and amend some of the work of Chenery referred to earlier. Chenery begins his pipeline analysis with the well known Weymouth formula:

$$Q = KD^{2.67}\sqrt{P_1^2-P_2^2} = KD^{2.67}P_1\sqrt{1-(1/R^2)} \tag{10}$$

The symbols have the same meaning here as in the previous section. A slightly different version of this formula is employed by C.H. Paulette (1968).

$$Q = 1434 \, D^{2.53} \left[P_1^2 - P_2^2\right]^{0.51} L^{-0.51} \tag{11}$$

Chenery standardises the distance between compressor stations to 100 miles (= 160 kilometers), which is also the figure used by Hawdon. It has been suggested that Paulette's equation permits calculating the optimal distance between stations, but I am not prepared to accept the result of such a calculation as other than a rough approximation: the optimal spacing of compressor stations depends on the optimal D and $p's$ and theoretically these things should be determined together.

Next Chenery introduces an engineering relationship for the maximum working pressure of a pipe p_1, with S as working stress (in pounds per square inch), and T' thickness.

$$p_1 = \frac{2ST'}{D} \tag{12}$$

This equation is then substituted in (10) to yield:

$$Q = K_1 D^{1.67} T' \sqrt{1 - (1/R^2)} \tag{13}$$

The opinion here is that this last operation was essentially without meaning. What equation (12) is intended for is determining the minimum thickness of pipe when we have the maximum pressure p_1, and D. Thus it is useful for determining the cost of the pipe, and its installation cost (which is a function of D, T' and L); but it is out of place in equation (10). For instance, as equation (13) now stands the suggestion is that production can be increased merely by increasing T', which is obviously not correct.

The other relationship that Chenery presents has to do with the required horsepower of a compressor station, given the compression ratio and the maximum amount of gas that the pipeline will carry.

$$H = (28.75R - 13.9)Q \tag{14}$$

The constants in this equation are determined from the thermodynamics of gases. With Q given, and equal to $\overline{Q}$, and with p_2 — the gas pressure at the compressor inlets — given, equations (10) and (14) suffice to construct an isoquant (i.e. a locus of constant population) in (H,D) space. This is so because with p_2 given, (10) can be solved to given combinations of D and p_1. Each p_1 (together with p_2) provides an H. Referring back to Figure 9.3 it can be appreciated that this simplicity is made possible by the assumption — pointed out by V. Smith (page 34) — that $p_0 = p_2$, or $R = p_1/p_2 = p_1/p_0$. We can also note here for future reference that various combinations of p_1 and D transform to T' via equation (12).

If we now look at equation (11), we observe that the length of the line is important for the amount of gas delivered. This is because of the pressure drop in the line that is caused by friction. This, in turn, suggests that the simplification employed in the above paragraph may not be completely justified, since there is no logical reason why p_2 should be equal to p_0: p_0 is independently determined by the pressure drop in the line between the reservoir and the first compressor station, while p_2 is determined by the drop in the line between stations. Accordingly, it might be edifying to view the above problem from a slightly different perspective. Following R.E.

Hodges (1985), this involves using three well-known equations from the thermodynamics of gases, and some empirical information provided by the American Petroleum Institute (API). Taking Q as given, these equations are:

$$V = \frac{60QT'Z}{D^2 p_1} = V(D, p_1) \tag{15}$$

$$f = \frac{0.021}{(\phi VD/u)^{0.38}} + 0.0025 = f(V, D) \tag{16}$$

$$\Delta p = \frac{f\phi LQ^2 T'Z}{(14.9) p_1 D^5} (1,000) = g(f, D, p_1) \tag{17}$$

In the above V is gas velocity, f a friction factor, Δp pressure drop, and L is the equivalent length between stations, which is given. The other symbols represent such things as gas density, gas viscosity, etc.

In observing these equations it should be clear that consecutive substitution yields $f = f(V(D, p_1), D) = f(D, p_1)$; and $\Delta p = g(f(D, p_1), D, p_1) = g(D, p_1)$. Next we can examine a short table involving the drop in pressure that was developed by the API.

Operating pressure (psig)	Acceptable pressure drop (psi/100 feet)
0 — 100	0.05 — 0.19
101 — 500	0.20 — 0.49
500 — 2,000	0.50 — 1.20

For arbitrary values of p_1, it is possible to set the acceptable pressure drop from the above table equal to Δp in equation (17), and solve for D. Accordingly it is once again possible to generate a locus in (p_1, D) space. Note that linear extrapolation can be used in this table: for example, a pressure of 1,000 psig corresponds to an acceptable pressure drop (per 100 feet) of 0.733. Once we have p_1 and the pressure at the entrance to the compressor p_2, we can use equation (14) once more to solve for H. Thus, once again, we have a point on an isoquant in (H,D) space. Hodges appears to have chosen p_1, and as a result H, on the basis of some independent criteria, since in his paper he is only concerned about D. On the other hand the computer program he developed provides information on all the variables in equations (15), (16) and (17), and also changes psig to psia.

We can now turn to the cost side of the analysis. For a given length of pipe, and assuming Q is fixed at $\overline{Q}$, we have C = C(W,H). (Q is fixed by negotiations between the buyers and sellers of gas. It is not determined by the intersection of MR and MC curves.) We also have W = W(T',D), and from equation (12) T' = T'(p_1,D); and so it is possible to write C = C(p_1,D,H). But p_1 = p_1(H,$\overline{Q}$), and so C = C(D,H). If we wanted to take into consideration some exogenous or autonomous costs, $\overline{C}$, this expression could be modified to C = C(D,H;$\overline{C}$). We can now set up the following cost-minimizing Lagrangian:

$$\hat{C} = C(D,H;\overline{C}) + \lambda \left[\overline{Q} - Q(D,H)\right]$$

Solving this is a simple matter:

$$\frac{\partial C}{\partial D} - \lambda \frac{\partial Q}{\partial D} = 0$$

$$\frac{\partial C}{\partial H} - \lambda \frac{\partial Q}{\partial H} = 0 \qquad (18)$$

$$\frac{\partial \hat{C}}{\partial \lambda} = \overline{Q} - Q(D,H) = 0$$

From this system we can solve for D and H, and subsequently for things such as the pipeline thickness T' and pressure p_1.

In Chenery's formulation, each of the cost variables H and D is associated with both fixed and variable components. For instance, the total compression cost is $(\alpha_1 i + \alpha_2)H$, where α_1 is the installation cost per horsepower of the compressor, and α_2 is the annual operating cost per horsepower, which includes the cost of the gas or electricity used to drive the compressor. i is what Chenery calls the combined annual rate of interest, depreciation and obsolescence — which makes it the same as the value of the annuity given in equation (9) above, at least in this simplified analysis.

To see this the reader should consider the following simple numerical exercise. Suppose we have an installation that costs 1,000 dollars, and will be used for two years. At the end of that time it can be sold for its original cost (or $C_T = C_2 = 1,000$ dollars). Some persons might say that the depreciation is zero, and in a sense this is true; but 1,000 dollars two years from now is not the same as 1,000 now. It is only $1,000/(1+r)^2 \approx 1,000/e^{-rT} = 1,000/e^{-2r}$ (and if r = 10 per cent is 826 dollars). In the general case, with C = C_T, we have for the present value of the asset:

$$C - \frac{C_T}{e^{rT}} = C\left[1 - \frac{1}{e^{rT}}\right] = C\left[\frac{e^{rT}-1}{e^{rT}}\right]$$

Using equation (9) to write this as an annuity we get for our yearly payment:

$$C\left[\frac{e^{rT}-1}{e^{rT}}\right]\frac{r}{1-e^{-rT}} = rC$$

This is the annual cost of a non-depreciable asset, and is due exclusively to the time value of money. Put another way, it is the opportunity cost of spending C on a physical instead of a financial asset. We also observe that in a situation where there is depreciation and obsolescence, and thus $C_T < C$, the annual charge for the asset (or capital cost as it should be designated) is larger than rC. Chenery took values for i of between six and ten per cent, which seems like a reasonable assumption considering that the real interest rate at the time he published his article was only about two or three per cent.

No attempt will be made here to obtain a solution to an actual problem, since I do not have access to any of the cost parameters that are needed to obtain such a solution; but Chenery has raised one issue that deserves some comment. This has to do with a pipeline where the load factor can become quite low. What this means is that the maximum capacity of the pipeline is employed only a small percentage of the time. To examine this problem let us consider a situation where there are only two choices: a great deal of compression and a relatively small pipe, as at $\overline{A}$ on activity A, and a large pipe with a fairly small amount of compression, as at $\overline{B}$ on activity B. It will first be assumed that at full capacity, $\overline{Q}$, both these activities have the same cost, and so if we always operated at full capacity we would be indifferent between the input combinations (H_A, D_A) and (H_B, D_B), which are shown in Figure 9.4.

Consider now the possibility of the load varying between $\overline{Q}$ and Q^*. As production declines below $\overline{Q}$, the decline in costs starting from A is greater than that of B by approximately $H_A H''/H_B H''$. This is because there are hardly any operating costs associated with pipelines, while lower load factors will mean a saving of energy used to drive the compressors, and appreciable savings might also be made where maintenance of the compressors is concerned. In fact, even if the cost of producing $\overline{Q}$ was lower at $\overline{B}$ than at $\overline{A}$, but the line was likely to operate at low load factors much of the time

Figure 9.4: Two production activities, A and B, for a pipeline

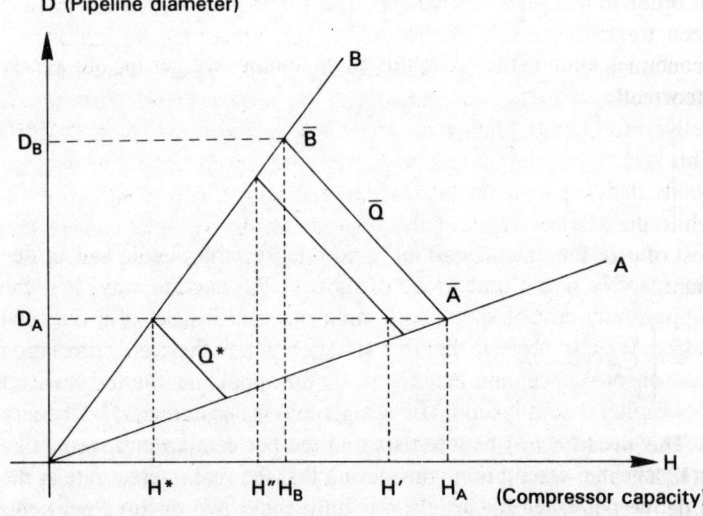

— for instance in the vicinity of Q^* — then the cost minimizing installation could be on process A at $\overline{A}$ (H_A, D_A). To see this let us suppose that $\overline{C}_A > \overline{C}_B$ when $Q = \overline{Q}$. Taking C_v as the variable cost, let us ask if there is a value of θ that will make:

$$\theta \overline{C}_B + (1-\theta)\left[\overline{C}_B - C_{vB}(\overline{Q} - Q^*)\right] > \\ \theta \overline{C}_A + (1-\theta)\left[\overline{C}_A - C_{vA}(\overline{Q} - Q^*)\right]$$

$(1 - \theta)$ is the portion of a period that Q is at Q^*, as compared with $\overline{Q}$, and clearly $0 \leq \theta \leq 1$. If this condition can be satisfied then $\overline{Q}$ should be produced at $\overline{A}$. From the above the implication is that we must have:

$$1 - \frac{\overline{C}_A - \overline{C}_B}{(C_{vA} - C_{vB})(\overline{Q} - Q^*)} > \theta$$

For instance, if $\overline{C}_A = \overline{C}_B$, then if $1 > \theta$ — or less than 100 per cent of the period is spent producing $Q = \overline{Q}$ — then production should take place at $\overline{A}$, using inputs (H_A, D_A).

AVERAGE AND MARGINAL COST

In order to complete the analysis I will touch on a matter that has been treated in much of the theoretical literature on natural gas economics, but to my way of thinking has usually been treated incorrectly. To start out, let us designate all variable costs from delivering Q units of gas per period over a time horizon T as wX. This is in, for example, dollars per period. The cost of all the capital goods that are used to deliver the gas will be called $p_K K = C_0$, while the salvage value of these goods at time T is C_T. The total cost of production, referred to the beginning of the initial period, is then:

$$C = \left[C_0(Q) - \frac{C_T(Q,T)}{e^{rT}} \right] + wX_0(Q) + \int_0^T wX(Q)e^{-rt}dt \quad (19)$$

The production function employed in this exposition is $Q = Q(K,X)$, and over the time horizon T an amount of gas equal to V is delivered, with $V = TQ$. It should also be obvious that with Q constant, an increase in T is equivalent to an increase in V. Furthermore, we get from differentiating (19) with respect to T:

$$\frac{\partial C}{\partial T} = - C_T(Q,T)e^{-rT}(-r) - e^{-rT}\frac{\partial C_T(Q,T)}{\partial T} + wX(Q)e^{-rT} > 0$$

Note here that $\partial C_T/\partial T$ is negative: the value of assets declines with time. Another differentiation gives:

$$\frac{\partial^2 C}{\partial T^2} = -r^2 C_T e^{-rT} + re^{-rT}\frac{\partial C_T}{\partial T} - (-re^{-rT}\frac{\partial C_T}{\partial T}) - e^{-rT}\frac{\partial^2 C_T}{\partial T^2}$$
$$- r(wX)e^{-rT}$$

The only term in this expression that has any ambiguity about it is $\partial^2 C_T/\partial T^2$, and to save time and effort in determining its sign I propose to make it zero. Thus we have $\partial^2 C/\partial T^2 < 0$. These last two results imply that as T (i.e. V) increases, with Q constant, C increases — but at a decreasing rate. Thus we have the arrangement shown in Figure 9.5(a).

Next we look at the matter of an increase in production when a given volume is to be produced. As proposed by A. Alchian (1959), producing a larger quantity per time period means, *ceteris paribus*, a higher total, average, and marginal cost. In conventional production theory, where we do not concern ourselves with volumes, this

Figure 9.5: Cost curves given changes in V and Q

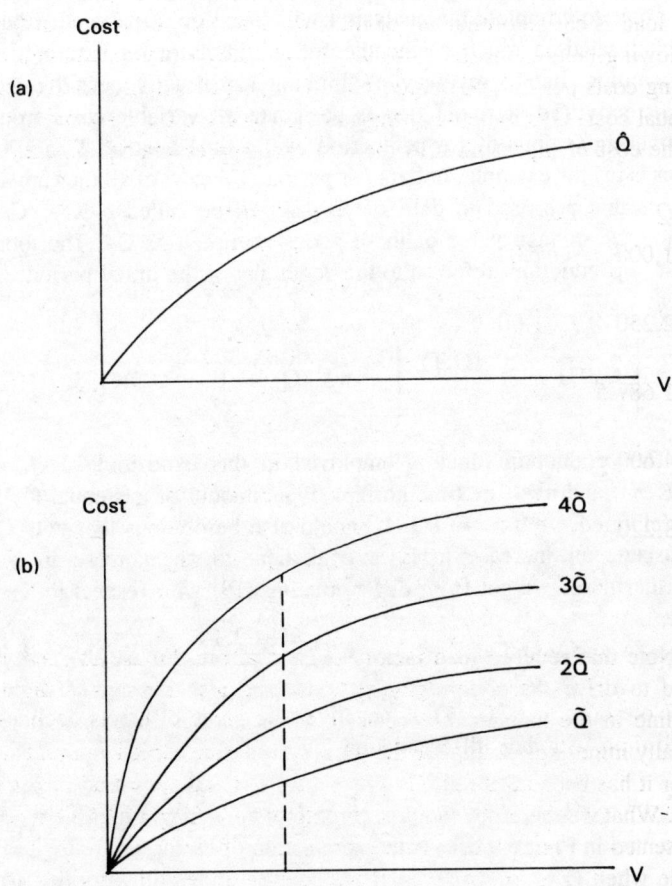

may not be true, however. Doubling production, for instance, can in some situations simply mean doubling costs. Alchian takes as a general proposition $\partial C/\partial Q > 0$, and $\partial^2 C/\partial Q^2 > 0$, but clearly this would not always be so in those cases where there are pronounced returns to scale — which is often true with pipelines. Instead I would expect that with V given, a rising Q would first result in increasing returns to scale, and then in decreasing returns to scale, where again one of the important causes of the decreasing returns to scale could be low load or capacity factors. To get some insight into this important issue, suppose we have a situation where at full capacity we have increased returns to scale: output per unit cost rises as production increases. At the same time let us postulate that half of each period the load is equal to one-half of full capacity. As can be seen in the following tabulation, the less-than-full capacity situation can cause rising costs per unit of output. Here Q signifies output and C total annual cost. G is the cost of the gas used to drive the compressors, P the cost of pipe; and C is the cost of compression equipment.

Q	Load factor	Annual cost	Q/C
1,000	1.00	1,000	1.0
		200 G; 400 P; 400 C	
2,250	1.00	2,000	1.125
		400 G; 800 P; 800 C	
1,687.5	0.75	1,900	0.888
		300 G; 800 P; 800 C	
4,600	1.00	4,000	1.15
		800 G; 1,600 P; 1,600 C	
3,220	0.75	3,800	0.847
		600 G; 1,600 P; 1,600 C	

Note that reduced load factors cause a fall in the use of energy used to drive the compressor stations, and also perhaps a slight decline in the maintenance costs of these stations; but since it is usually impossible to reduce the cost of expensive capital equipment after it has been acquired, C/Q rises (i.e. Q/C falls) as load factors fall. What we therefore should expect to see in a diagram of the type presented in Figure 9.5(b) is the domination of increasing returns to scale when Q is low — even if we operate under full capacity an appreciable portion of the time; but as Q gets larger the influence of low load factors (i.e. unused capacity) on the unit cost becomes more pronounced, and we get decreasing returns to scale.

Using equation (19), let us examine dC/dQ, with T constant and adding the simplification that the pipeline and compressors have no scrap value at time T. In other words, $C_T = 0$. In this operation we are dealing with a total derivative, since both Q and V will be allowed to vary. Indeed, these are fixed by the relationship $V = TQ$.

$$\frac{dC}{dQ} = \frac{\partial C_0}{\partial Q} + w \frac{\partial X_0}{\partial Q} + \int_0^T w \frac{\partial X}{\partial Q} e^{-rt} dt$$

$$= \frac{\partial C_0}{\partial Q} + w \frac{\partial X_0}{\partial Q} + w \frac{\partial X}{\partial Q} \left[\frac{(1-e^{-rT})}{r} \right]$$

The first two terms on the right hand side of the above expression are $\partial C/\partial Q$. They represent cost changes associated with changes in output of the type shown in Figure 9.5(b). To see the significance of the last term on the right hand side let us write it in terms of finite increments, and multiply by $\Delta V/\Delta V$. Then we get:

$$w \frac{\partial X}{\partial Q} \left[\frac{(1-e^{-rT})}{r} \right] \approx w \frac{\Delta X}{\Delta V} \left[\frac{(1-e^{-rT})}{r} \right] \frac{\Delta V}{\Delta Q} \approx \frac{\partial C}{\partial V} \frac{\partial V}{\partial Q}$$

The interpretation of $\partial C/\partial V$ is as follows: it is the total discounted value, over a time horizon T (excluding $t = 0$), of the change in variable costs caused by a change in production. Marginal cost thus becomes, when the effect of volume is taken into consideration:

$$\frac{dC}{dQ} = \frac{\partial C}{\partial Q} + \frac{\partial C}{\partial V} \frac{\partial V}{\partial Q} \tag{20}$$

It might now be useful to draw the following diagrams (Figure 9.6).

The average cost (AC) and marginal cost (MC) curves are calculated from the locus OL, and are valid for a given time horizon T. Changing T means changing the locus OL.

Figure 9.6: Marginal and average cost curves when volume is explicitly considered

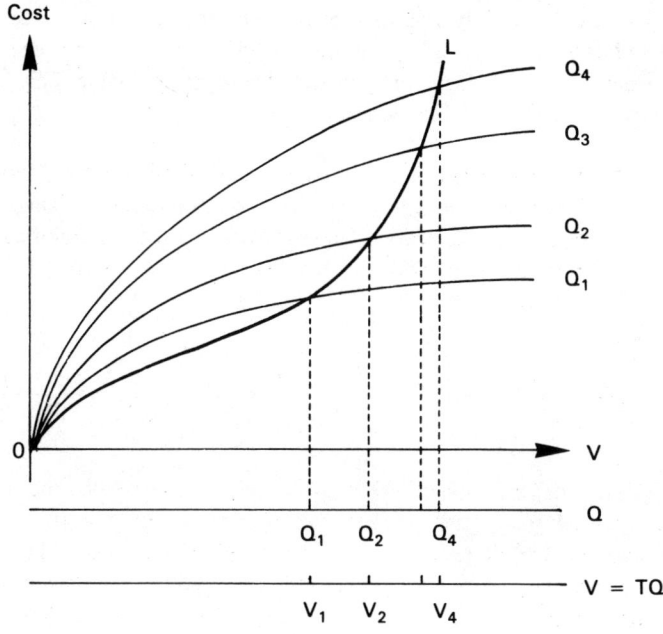

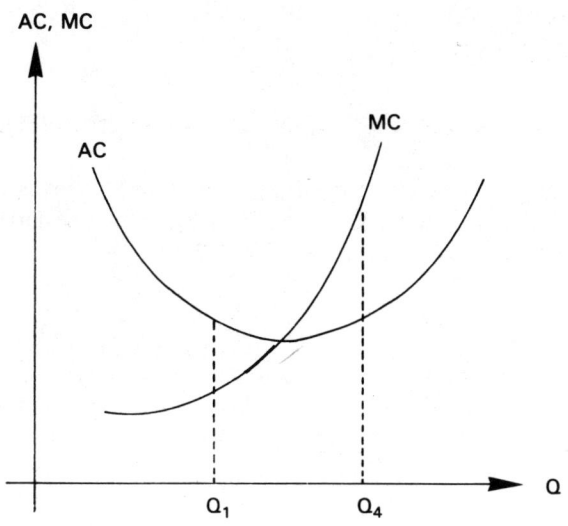

CONCLUSION

There is still a great deal of work to be done on this important topic. I find it only slightly less than astonishing that in the past decade there has not been a single paper on pipelines in the academic energy journals, although from time to time there are valuable technical notes in the *Oil and Gas Journal*.

10

Energy and the Macroeconomy

This chapter provides a brief examination of three important topics in energy economics: the relation between energy and the macroeconomy; futures markets; and the long term supply of oil. Over a period of only a few months — from the end of 1985 to March 1986 — the spot price of oil fell from about 30 dollars/barrel to 14 dollars/barrel, and occasionally dipped below this level. In fact 'future' oil has sold for an average of about 12 dollars/barrel, and on occasion has fallen below 10 dollars/barrel. It is reasonable to expect a slight recovery in the price of oil, and expectations are that the price will eventually stabilise somewhere between 15 and 20 dollars/barrel, however it is generally believed unlikely that the oil price will regain its 1984–5 level during the present decade. The world has, at least temporarily, been returned to an era of inexpensive oil, and the main question now is what do we have a right to expect from this unexpected but delightful bounty.

First we must ask if the markets for goods, labour and financial assets will now function in such a way as to restore the world economy to the condition that we once referred to as normal. The opinion here, which will be elaborated on below, is that there will be no reverse oil shock in the sense of a 180 degree change from the previous situation, at least not in the short run. This is because several macroeconomic markets are still out of equilibrium, and it will take a long time for them to adjust. Even so, the present situation is favourable for the oil-importing countries. Taking just the industrial world, it appears that a fall in the oil price creates many more winners than losers, and the same is probably true for the Third World. The losers, at least in the short run, will be the oil-exporting countries and, as will be explained toward the end of this chapter, some of these countries might also be losers in the long run:

as things now stand the oil revenues of the 13 OPEC countries were likely to fall below 50 billion dollars in 1986, which is the lowest level since 1973, when they were 23 billion dollars. This is a huge difference from the 158 billion dollars these countries earned in 1984, which was also a year when the value of the dollar was in the process of climbing to record heights *vis-à-vis* the currencies of the other industrial countries, which meant that the actual purchasing power or terms-of-trade of a barrel of oil was higher than ever.

By way of discussing the goods market, let us consider the situation in the United States. It has been calculated that a fall in the oil price to between 15 and 20 dollars/barrel might eventually add one per cent to the inflation-adjusted Gross National Product (GNP). For Japan and Western Europe the effect should be greater, since these areas are more dependent than the United States on imported energy. Forecasts of the OECD, IMF and similar organisations suggest that the economic growth of the industrial world will be at least three per cent/year for the next few years; and if this is true, many large industries are going to be summarily pulled out of the doldrums. Among these should be chemicals, steel, airlines, cars and independent oil refineries.

Chemicals will gain to the extent that they are heavily involved with petroleum-based feedstocks (as, for example, naphtha), which in some cases account for 25 to 50 per cent of operating costs; but even those chemical firms that use natural gas should do better, because the price of natural gas will tend to decline in phase with oil. Steel (and other energy-intensive industries such as copper and aluminium refining) will gain for virtually the same reason; oil is now only a small percentage of the steel industry's energy input, but all energy prices are, at present, linked in some way to the price of oil. In addition, the economic position of steel buyers would be improved, and this would tend to raise the demand for steel and steel products. Airlines will gain from cheaper fuel in an obvious and straightforward way. The same is true for the car industry: there will be large financial savings from lower cost energy, and in addition the fall in the cost of using cars and other vehicles will increase the demand for these items, while the decrease in the energy cost of producing them should help to prevent a rise in their price. Finally, the many independent oil refineries in the United States, Western Europe and Japan will obtain less expensive raw materials, which in many cases will probably mean both higher profits and cheaper petroleum products — such as gasoline — which will tend to raise the demand for these products.

One of the markets that has not adjusted as yet is the labour market. In most industrial countries (and of course in the Third World), the unemployment rate is now very much higher than it was before the first oil price shock, and it seems unlikely that the levels of the 1960s and early 1970s can be restored. This is because some fundamental and, to a certain extent, undesirable changes have taken place in the structure of production in many industrial countries; and because of the increase in international competition due to accelerated economic development in the newly industrialising countries. This accelerated development was financed to a considerable extent by petrodollars (i.e. revenues of the oil-exporting countries that were recycled to financial institutions in the industrial world, where they were initially loaned to the newly industrialised countries and the industrial countries at low real rates of interest). In an economic sense the rise of these newly industrialising countries (Singapore, Taiwan, etc.) is an anologue of the political changes that took place in the former colonial possessions of Britain, France and Holland due to the military successes of the Japanese in 1941–2: these successes eventually upset the balance of political power, just as the rise of OPEC has permanently upset the balance of economic power.

Returning to the labour market, in the United States in 1983 after a decision was taken to abandon the more counterproductive aspects of monetarism, a business cycle upswing began that was the strongest in the industrial world. Even so the share of employment in the manufacturing sector continued to shrink, and has fallen even faster than in Europe: from January to June of 1985 alone, 160,000 manufacturing jobs were lost as the industrial sector continued its long decline, even though the aggregate rate of growth of the United States economy had been the highest in the OECD over the previous year. In the past decade the most dynamic expansion in the United States has taken place in the service sector, where almost 12 million jobs have been created, with approximately 600,000 of these positions coming into existence in the first 6 months of 1985. Most of the new jobs are in eating, drinking and health care, where wages are on average 20 per cent below corresponding levels in manufacturing. This reflects the fact that the value of output/employee in manufacturing has tended to be considerably higher than for persons employed in the service sector, and also higher than for persons employed in trade.

Accordingly, while the employment gains are admirable, it may be that economic progress in the United States is beginning to assume a different character from that prevailing earlier in this century, to include a greater susceptibility to the condition that Keynes described as a deficiency in effective demand. This is true because even now the manufacturing sector produces 50 per cent of the gross national product of the United States, although it employs only 20 per cent of the non-agricultural work force; and thus if manufacturing continues the decline that began in 1980, the demand for services by firms and the employees of firms in this sector must also decline. For this not to happen, employment in the service sector must expand much more rapidly than employment falls in the manufacturing sector. (Traditionally service jobs were created because of the expansion of output and job opportunities in manufacturing). Although unemployment has greatly increased over the past half decade, the service sector has thus far been able to create enough new jobs to maintain demand, but many of us do not believe that this situation can continue. Instead, if the economy slows down, fewer service-related positions will be created, and before the policymakers can raise effective demand, unemployment will begin to accelerate and demand will fall even faster.

Things may not be improved by a rise in productivity in the service sector, because if fewer people can produce the same amount of goods, there could be a reduction in the demand for labour by that sector. In the United States there has been a very large surge in technological investment per service worker — from 415 dollars in 1975 to nearly 1,000 dollars in 1985 — and the result is showing up in a stagnation in employment growth in those service areas where rapid technological upgrading is possible. Initially this will be compensated for by the equipment purchases of the service sector, but this cannot last indefinitely.

In the United States, and perhaps the entire OECD, many new jobs are being created in the area known as high tech, and there is even some talk of the present industrial world finding its salvation in the design and manufacture of robots, computers and so on. I depart company from this kind of thinking, and advise my more sensible colleagues to do likewise. On the average, high-tech industries require even less capital and technology than heavy manufacturing, and as a result they could gravitate to the newly industralising countries (NICs) with the same ease as steel and shipbuilding. Yes, the output of the high tech sector will grow fast compared with other sectors, but because the high tech sector is small, and its productivity is increasing rapidly, it will only be capable of creating a small number of jobs. What it all comes down to is that — directly or indirectly — the de-emphasis on manufacturing will shuttle larger numbers of people in the industrial world into low-tech occupations or, if they are unlucky, unemployment. According to my interpretation of income statistics, this process is in full swing in several countries, and the United States seems to be one of them.

THE INTERNATIONAL ECONOMY AND FINANCIAL MARKET DISEQUILIBRIA

An important matter that was not explained above is the relationship between the fall in the oil price and the value of the dollar. Simple observation has shown that increases in the price of oil appreciate the currencies of energy-rich countries relative to those of energy-poor countries. One reason for this is that after an energy price rise, energy-poor countries have to sell more goods in order to maintain their import of energy materials, and one way to increase the attractiveness of these goods is to lower their price in terms of foreign currency: i.e. depreciate their currency. There is also the situation that oil is paid for in dollars. If the price of oil rises, the energy-poor countries must increase their demand for dollars (or, equivalently, increase the supply of their own currencies). Again, this tends to appreciate the dollar relative to their currencies. With the oil price falling we should expect to see things moving in the other direction, and this is what has been happening: from the peak it reached in February 1985, the dollar fell by 30 per cent against the yen and Deutschmark before the end of February 1986.

For reasons that can only be touched on here, the fall in the value of the dollar has been something that was greatly desired, even by the countries that have been benefiting by the decreased competitiveness of United States goods and services that resulted from the overvalued dollar. The reason is that there has been an unhealthy accumulation of dollars outside the United States because imports into the US were stimulated by the low exchange value of foreign currencies and the huge United States budget deficit, much of which was financed by foreign borrowing. This can also be described as a disequilibrium, and potentially an extremely serious situation for everybody. With United States foreign debts growing faster than income, the day would eventually come when foreigners concluded that it was no longer sane to continue to accumulate dollars (or dollar assets), and the value of the United States dollar would have to fall.

The question at that point would be how fast it fell, because if it crashed the experience might be traumatic for the international financial system. It would become more difficult, or even impossible, for the United States to service its huge international debt (which would heap even more suspicion on the dollar, and cause it to fall faster unless the United States economy was deflated at an unhealthy rate); and foreign businesses, persons and financial institutions holding United States financial assets (to include central banks) would find their assets depleted — which could mean a dramatic fall in demand, and most likely recession, in these creditor countries.

Because of the declining oil price this may not happen. If the dollar depreciates slowly, many of the dollars held abroad will return to the United States in an orderly fashion as part of the demand for United States goods and services. Some observers are not optimistic about this sort of activity taking place on a scale that is large enough to shave the mountain of dollar assets held in Western Europe and Japan to a tolerable level; however if the United States budget deficit can be brought under control, and things seem to be moving in this direction, I see no reason why, in a world where trade continues to expand, and the dollar slowly to depreciate, the United States external debt cannot be stabilised at an acceptable level in the present decade.

Excessive private debt in the leading industrial countries, and especially the United States, is capable of posing a major threat to the international economy. In the United States private debt has now reached a record level in terms of the US gross national product. Much of this debt is due to investors borrowing in order to purchase financial assets that are increasing in value because the fall in energy prices causes a substantial appreciation in their present value. For example, in line with the discussion above, a fall (or even an expected fall) in the price of energy would greatly improve the profitability of the steel and car industries, and sooner or later this should be reflected in their share (i.e. stock) prices. Since 1985, share prices have surpassed all previous peaks on all except two of the largest stockmarkets in the world (Singapore and Hong Kong). The stockmarket values of many companies have now reached a point, however, where they resemble the actual value of these firms (in terms of profits that will be eventually realised) only if the world economy continues to expand at its present rate. Otherwise this market is also out of equilibrium, and the adjustment here might resemble that of 1929. At that time, according to many economic historians, the principal difficulty was that too much stock was bought on 'margin' — that is to say, using borrowed money. Accordingly, when share prices began to fall as the adjustment to authentic asset values began, more margin was requested (i.e. a small percentage of the total outstanding loans were called in), but now regiments of borrowers could find only a few platoons of lenders, and so these borrowers attempted to obtain this margin by selling some of their shares. The result was that the market fell even faster, and they had to sell more and more. It was thus that the Great Depression began.

FUTURES MARKETS

Futures markets operate as follows. Against a background of speculators 'betting' on the direction and size of commodity price movements by buying and selling futures contracts, an impersonal agency is created which permits producers, consumers, inventory holders, and other traders in physical products to reduce (i.e. hedge) undesired price risk. In the case of crude oil and oil products, futures markets promote a less ambiguous setting of prices, and a general upgrading in the efficiency of physical markets for oil and oil products because more information is available to existing and potential market participants. It has been said that futures markets help reduce erratic price movements, but I am not sure about this: inventory demand is strongly influenced by changes in expectations about future prices, and any institution which encourages frequent revisions of expectations can be identified as a potential source of increased price volatility.

The success of a futures market depends on its satisfying several criteria: the commodity should be traded in bulk and homogeneous, although different grades can be traded at a premium or a discount; production and consumption should be widely distributed, so as to preclude a 'corner' on the market; trade should take place at an exchange organised as an auction market; and the physical commodity should be bought and sold in circumstances that cause its price to fluctuate in a random or non-systematic manner. Without this latter provision speculators will not be attracted to the commodity, and without fairly large-scale speculation futures markets will not function properly. Traders in a physical commodity can employ futures markets to reduce price risk only if other traders (and/or speculators) are willing to accept this risk. The social gain from futures trading derives from the voluntary redistribution of risk between speculators and risk-averse dealers in physical products.

Now we can turn to the *modus operandi* of speculators. If a speculator believes that the price of a commodity is going to rise, he buys futures contracts. If the price of the physical commodity does rise, then for reasons discussed below, the price of the futures contract will also rise, and by selling the contract the speculator will make a profit. Moreover, in a viable futures market it will always be possible to sell this contract. The thing to be appreciated is that futures contracts are forward contracts, in that conditions are stipulated on them relating to the delivery of a specified amount of a commodity to a specified location or locations; but it is possible to avoid taking delivery if, at any time before the contract matures, an offsetting sale is made of a contract for the same amount of the same commodity, referred to the same delivery month. Offsetting is generally referred to as closing a position. Similarly, if a speculator thinks that the price is going to fall, then he opens his position by selling a contract, hoping to make a profit by closing his position at a lower price.

Hedgers also buy and sell futures contracts, depending upon whether they want to guard against price rises or price falls. Consider, for example, a refiner who has contracted for a given quantity of oil, but does not know the price at which this oil will be delivered because the supplier charges the spot price of oil at the time of delivery. This buyer thus faces considerable price risk because the price of oil might rise sharply. Risk-averting buyers can lock in the price at which they will receive their supplies if they buy futures contracts at the time they contract for their oil, while making an offsetting sale around the time the oil is delivered. If the spot price of oil rises the buyer takes a loss on the physical transaction; but since the price of futures contracts should rise in phase with the rise in the price of the physical commodity, a compensating gain can be made on the sale of figures. A perfect hedge is when the loss on the physical transaction is equal to the gain on the paper transaction — or vice versa. Perfect hedges rarely happen on individual transactions, but over a large number of transactions the small gains and losses per transaction should balance out.

Recently, and perhaps for the first time, the large oil firms have become active in the futures markets. These oil companies have control over a certain amount of oil during a given period, and if this oil is not contracted for by their customers, it is sold on spot markets where *ex ante* prices are unpredictable. But if these firms sell futures contracts and the price of the commodity they are selling falls, then if the price of the futures contracts also falls (which must happen sooner or later) an offsetting purchase of contracts can be made at a lower price than the selling price. Consequently the profit realised on the paper transaction tends to counterbalance the loss in the physical (or actuals) market.

We now come to the crucial matter of the convergence of the price on the futures market (the futures price), and the spot (cash) price of the physical commodity. It is only when futures and spot prices converge that a futures market can provide hedgers with a means for avoiding price risk. This should always be the case in unconstrained markets where the seller of a futures contract has the right to make delivery on the contract; and the buyer of a contract can hold open the contract, and obtain delivery. If, for example, the price of a commodity in the spot market is higher than the price of the commodity on a futures contract at the time the contract matures (e.g. at the close of the last trading day of the month before the delivery month), then someone who bought a futures contract would not make an offsetting sale of a futures contract. Instead, he would accept delivery of the commodity and immediately sell on the spot market. This kind of arbitrage should soon eliminate the price discrepancy. Similarly, if the price of a commodity on the spot market is less than the futures price at the maturity of the futures contract, the seller of a futures contract purchases the commodity on the spot market (rather than making an offsetting purchase of a futures contract), and exercises his right to deliver the commodity. Once again, this type of arbitrage should reduce the discrepancy between the prices in both the futures and physical markets.

I will now construct an artificial but pedagogically useful example of a futures market with three traders — speculators X, Y and Z — who are involved in transactions on three trading days. This example works because I am overlooking the pivotal factors margin and marking-to-the-market, which are essential features of any real futures market. When delivery is discussed, it will take place at the prices on the contracts. By way of contrast, the practice on the important New York Merchantile Exchange (NYMEX) is to determine the delivery price by the settlement price for the contract on the futures exchange on the last day of trade for that contract before the delivery month, and this is almost certainly true on other exchanges. Later I will argue, briefly, that the two specifications of the settlement price amount to the same thing because of the daily adjusting of profits and losses for or against the accounts of traders. Delivery takes place in only a minority of cases — conventionally only about one per cent, although somewhat more for oil and oil products.

The example is now presented, beginning with the no-delivery situation. In the following tables I list the trading day; price P at which transactions (i.e. sales of purchases of futures contracts) will take place in dollars/barrel; the volume of transactions V; and open interest OI. Open interest, which is important, is defined as the number of open (i.e. not closed) contracts, bought or sold, at the end of a given period — but not both. The contract being used in the illustration is for delivery in December, with transactions taking place in November. All transactions are for one contract — 1,000 barrels of crude oil.

Day	Price	X	Y	Z	V	OI
1	28	Buy (long)	Sell (short)	—	2	1
10	27	—	Buy (long)	Sell (short)	2	1

In the above, the purchase or sale of a contract is considered a transaction, and so there are two transactions on each day. Open interest at the end of day one is 1. This is because open interest represents the number of contracts that have been bought or sold during a given period, but not both, and not closed out. In this example I will only count purchases. Let us assume that there is no trading until day ten. At the end of day ten, Y closes out his position with a purchase that offsets his previous sale, while Z opens a position. But open interest, considered in terms of the transactions taking place on day ten, is still 1: this is because, since we are counting only purchases, the only contract that interests us is the one held by X, and which is still open. Up to that point in the month open interest is 10. Observe also the expressions long and short. Long means that the commodity is to be delivered to the holder of the contract, while short means that the holder of the contract must deliver the commodity (unless, of course, the contract is closed). On day ten trader Y realises a profit of one dollar/barrel on his transaction: he is therefore paid 1,000 dollars by the clearing house. Now let us assume that there is no more trading until day 29, which will also be taken as the last trading day of the month. Then we have:

Day	Price	X	Y	Z	V	OI
29	30	Sell (short)		Buy (long)	2	0

On day 29 all contracts have been closed out: X and Z close out their contracts by a sale and a purchase involving each other. Open interest at the end of day 29 is zero, although volume on that day was two. Open interest *for the month* was 28. (Monthly statistics are often presented with the dynamic expansion of e.g. the New York Merchantile Exchange; end-of-day open interest is also often cited.)

X makes a profit of two dollars/barrel due to selling at 30, while Z makes a loss of three dollars/barrel, since he opened his position by selling at 27, and closed it by buying at 30. X receives 2,000 dollars from the clearing house while Z pays 3,000 dollars. The clearing house makes a profit of 1,000 dollars, but this profit exactly covers the loss sustained when it paid 1,000 dollars to Y on the tenth day of trading. As is easily verified, the clearing house always breaks even because of the double entry nature of each transaction. This situation can be generalised and results in the following identity:

$$(P^s_{29} - P^b_1)_X + (P^s_1 - P^b_{10})_Y + (P^s_{10} - P^b_{29})_Z \equiv 0$$

Let us see what happens if these traders take (and make) delivery, looking first at the situation where the delivery price is the price at which they originally bought or sold a contract. What delivery means is that they do not close out their contracts on day 29, with X — who originally bought a contract — taking delivery; and Z — who originally sold a contract — making delivery. Thus Z must buy 1,000 barrels of crude on the spot market, or take it from his stocks and deliver it to a location or locations specified in the contract. Proof of delivery is then given the clearing house, which pays Z 27 dollars/barrel. On the other hand X takes delivery at a specific location, and after being given proof of delivery, pays the clearing house 28 dollars for each of the 1,000 barrels received. The clearing house directly recoups the 1,000 dollars it paid Y earlier, since (28–27) × 1,000 is the profit of the clearing house on these two transactions. Note that in this example the spot price of oil is irrelevant. Whether Z paid 30 dollars/barrel for the oil, and/or X sold oil for that amount, is not of interest.

Now let us consider the arrangement mentioned above, in which the settlement price is the price of a futures contract on the last trading day of the month. What will be shown is that, in effect, this situation is precisely the same as the one discussed above. But before doing this, something must be said about the margin. For each contract both buyer and seller must post a small percentage of the nominal value of the contract with their broker. Contracts are then revalued (i.e. marked-to-the-market) each day at the market's closing prices. If the prices have moved below the previous day's closing prices, traders with long positions are asked to pay additional margin, while those with short positions (i.e. those who have sold contracts) show a profit which can be taken in cash without closing out the transaction. This process ensures that profits and losses are not carried too far forward. Many potential hedgers have been frightened away from the futures market by repeated margin calls, but the thing to be appreciated here is that if they are losing money on their futures contracts, they must be gaining on their physical transactions in one sense or another.

For the purpose of this next example the assumption will be that the margin is zero, and that any calls on the trader for more cash, or for that matter a remitting of profits to traders, takes place on the last trading day of the month (instead of every day through the month). Then, in terms of the above example Z owes the clearing house 3,000 dollars, which are losses on his initial position, while X should receive 2,000 dollars, which are X's gains on his initial position. The result is a 1,000 dollar 'profit' to the clearing house, and this is the same as above. (I am assuming that the traders are dealing directly with the clearing house instead of through their brokers.) In addition, since Z is delivering oil, he will receive 30,000 dollars from the clearing house, but the clearing house will be receiving 30,000 dollars from X who is taking delivery. The total balance for the clearing house is thus zero once again, since the 1,000 dollar profit goes to cover the previous loss of 1,000 dollars on Y's transaction.

At trading time on the 29th day, X has a contract worth 30,000 dollars (30 dollars/barrel). Even if he can buy oil for 28 dollars/barrel — or 28,000 dollars for the contract — which was the price of the contract on day one, he is losing the 30,000 he could gain by selling the contract (assuming that he has not received any of his 2,000 dollar profit earlier). Thus, presumably, if he takes the oil instead of selling the contract, the oil is worth at least 30,000 dollars to him. Looked at from this point of view it does not make any difference whether the futures contract was a forward contract, and oil could be bought for 28 dollars/barrel — or if X had to pay 30 dollars/barrel for oil after receiving two dollars/barrel in profits. In fact, if the mark-to-the market procedure that was mentioned above is in effect, then the true settlement price must be 30 dollars/barrel.

I close this section with a few remarks that the reader should keep in mind when pondering the future of futures markets, to include proposed futures markets for natural gas. All futures exchanges are membership organisations, in that members are individuals or corporations that buy their seat on the exchange. These members can trade on their own account or for customers. Many exchange members seem primarily concerned with the former and can be classified as small speculators. They are not expanding the liquidity of the exchanges by a large amount (which is desirable); instead they are tending to increase pit congestion to a point where the face-to-face, open outcry and hand signal system that has prevailed in these markets for hundreds of years may have to be abandoned. In New York these new speculators trade mostly with their own money, and mainly think in terms of closing all their positions at the end of the trading day with a small profit. This is one of the reasons why open interest in oil futures has tended to develop so sluggishly.

It has been suggested that one of the answers to pit congestion is computer trading, and without doubt this is an idea whose time has come. However even wall-to-wall computerisation will not create the liquidity that is required if the larger oil companies or the oil-exporting countries in OPEC use these exchanges to the extent that some people feel that they should be using them. In many respects these exchanges are still relatively small affairs, and cannot be anything else in the near future, although the trade in 'paper' barrels seems to be setting new records every week. The reason for this is that in actual, as compared to theoretical, futures markets, it is impossible to provide a wide enough variety of contracts to insure against all sorts of price risk. In fact it may be true that United States futures markets offer potential hedging coverage for only 20 per cent of the oil products being produced. Furthermore, since inventory behaviour is influenced by changes in expectations about future prices, institutions such as futures markets which encourage frequent revisions of expectations can be identified as a potential source of increased price volatility. For these, and other reasons, the popularity of futures markets among many potential hedgers and/or producers and traders in physical products has never matched the popularity of futures markets with speculators, at least where oil is concerned.

CONCLUSION

I will conclude this chapter, and book, with some remarks on the world oil market (which, as stated earlier, sets the pace for all energy markets).

By 1985 the output of OPEC crude oil was 16 million barrels of oil per day (= 16 Mbbl/d), which was half its 1979 production, and less than 60 per cent of capacity. OPEC's share of the world market outside the Centrally Planned Economies was 38 per cent, compared to 60 per cent in 1979. The largest part of this reduction was absorbed by Saudi Arabia, whose output fell not only in phase with the total decline in OPEC output, but because that country had, for years, voluntarily reduced its output in order to offset above-quota production elsewhere. Eventually the Saudi Arabian government reached the conclusion that there was no economic or — given the independent pricing and sales practices of certain OPEC members — moral reason to continue to function as OPEC's 'swing' producer. It was made clear that Saudi Arabia would, in the future, produce and sell *at least* its assigned quota of about 4.3 Mbbl/d. As a result the traditional policy of fixed crude oil prices was abandoned in favour of a system of flexible prices determined on a netback basis by the market demand for refined petroleum prices. In other words, given the market prices for these refined products, Saudi Arabia declared to various purchasers of crude that it was willing to sell oil to these purchasers at such a price that it was always profitable to buy Saudi oil.

This is bad news for most of the other oil exporting countries because with the exception of Kuwait and the United Arab Emirates — which are cooperating with Saudi Arabia, and have also expressed an interest in netback deals — Saudi Arabian oil is the least expensive in the world. This means that if these other oil exporters want to deny the Saudis their new self-imposed quota, some of them will undoubtedly find themselves selling oil for less than its cost of production. If we think in terms of the total cost of production (i.e. both fixed and variable costs), things have already reached that stage for several major oil producers; but if we think only in terms of variable costs (which are perhaps the relevant costs in this situation), then the price of oil will have to fall to under eight dollars/barrel, and perhaps even lower.

Saudi Arabia intended to sell at least its full quota at the official price of 28 dollars/barrel, because the Saudis estimated that this quantity at this price satisfied their revenue requirements. But, at least in theory, if the price fell they were prepared to sell whatever quantity was necessary at whatever price was necessary in order to obtain the money they felt they needed — although, in practice they were prepared to, and have, taken external political constraints into consideration. Given the huge amount of low-cost reserves found in the core countries of the Middle East (Saudi Arabia, Kuwait, the UAE), the other oil-producing countries might find themselves facing an extremely long period of low oil prices. Many of these countries are unable, for both economic and political reasons, to cut production and wait for this period of low prices to pass, but by the same token they cannot afford to produce at prices beneath the level of those prevailing at present (which, in March of 1987, seem to be hovering at around 17 dollars/barrel). But they are producing anyway. Thus the rest of the world is at present consuming irreplaceable oil which should and perhaps could have been selling for 25 dollars/barrel, at prices that have dipped to half this amount.

Next we can ask if this is an ideal situation for the OPEC core countries, apart from the bad feelings that have been raised elsewhere in the oil-producing world. Obviously, even if these OPEC core countries possessed hundreds of years of oil, they also prefer to sell at high rather than bargain basement prices; but although their reserves are huge, they are not unlimited, and the governments of these core countries *are* concerned about future generations: unlike the governments of many industrial countries, their actions make this clear. Their optimal strategy must therefore be to sell oil at a low price until those members of OPEC who had been experimenting with independent pricing strategies fully comprehend that this was a mistake, and adjust their output in such a way that the money price of oil begins to move toward its previous level. Later, when the oil market is once again back in balance, it will probably be possible to start raising the real price of oil (which has been falling because of the gradual slide in the money price, inflation in the oil-importing countries from which OPEC obtains its imports, and the depreciation of the dollar). But the ability of OPEC to raise the real price of oil also depends on the reserve situation outside of OPEC: when oil was about 30 dollars/barrel, oil reserves were falling very rapidly in the British North Sea and the United States; but if these reserves do not continue to fall, they will be available to dampen future price rises. Thus an important component of an optimal strategy must be to set a price that will continue to encourage the depletion of reserves outside of OPEC.

What we have in the above are contradictory goals: low prices on the one hand, which enable the core countries to satisfy their revenue requirements, and which at the same time discourage exploration and substitution; but on the other hand high prices that will lead the countries outside of OPEC to continue producing at such a level that they cannot exert a strong influence on oil prices in the future. It is not clear what strategy, if any, the oil-importing countries are planning in order to maintain their present edge. It is possible, however, that this edge will last only a few more years, and during this period everything possible must be done to prepare the industrial world for the possible return of high-priced energy. In particular steps must be taken to reduce the present level of unemployment, which is the number one political, philosophical and economic problem of our time.

A basis for cooperation between oil-producing and oil-importing countries must be found. Viewed in terms of generations, rather than years, there is too much for everybody to lose to permit the unleashing of an oil cycle distinguished by long periods of oil prices that are either too high or too low and which, contrary to a growing belief, make it impossible for any type of market other than a casino to operate efficiently.

APPENDIX: SHORT-RUN OIL PRICING AND THE RESERVE/PRODUCTION RATIO

It has been suggestedd that in the short-run oil producers can gain by decreasing their prices. Some producers have begun to act as though this is true, although elementary economic theory suggests that it is false. Suppose for example that there is a ten per cent decline in demand. If prices remain unchanged then it can be shown that the decline in revenue R is also ten per cent.

$$R = pq$$
$$\Delta R = p\Delta q + q\Delta p$$
$$\frac{\Delta R}{R} = \frac{p\Delta q}{pq} = \frac{\Delta q}{q} \quad \text{(if)} \quad \Delta p = 0$$

Here p is price, q quantity and R revenue. Going back to the original assumption that demand declined by ten per cent we get:

$$\frac{\Delta q}{q} = -10\% \rightarrow \frac{\Delta R}{R} = -10\%$$

But suppose the intention is to maintain output. If the short-run price elasticity of demand is -0.5, which is probably high, then to raise q by ten per cent (after its initial fall) implies that the price must be lowered by:

$$\frac{\Delta p}{p} = \frac{1}{n}\frac{\Delta q}{q} = \frac{1}{-0.5} 10 = -20\%$$

Here n is the (short-run) price elasticity, and so revenue change is:

$$\frac{\Delta R}{R} = \frac{\Delta p}{p} = -20\%$$

This is so since the original change in demand was 'reversed' by the price fall, and thus the total change in q is zero. If the price elasticity is less than (minus) one, in the short run it is irrational to attempt to maintain revenue by decreasing the price. Of course this does not mean that in the short run some oil producing countries cannot make gains at the expense of others, but these gains are definitely temporary. The dilemma is that there may be gains for all the oil producers in the medium to long run if oil prices come down. I suspect in fact that at the present time (mid-1986) there is probably an oil price 'somewhere' in the vicinity of 20 dollars/barrel, perhaps slightly lower, that is optimal for everybody — both producers and consumers; but at the same time I doubt whether this price can be sustained in the face of the present tendency towards piecemeal and/or *ad-hoc* adjustments of the price by individual producers. (And, if these tendencies continue, the same is true of any price, regardless of how low it is.)

Reserve/production ratio

In my lecture at the 1986 International meeting of the IAEE, I stated that in the coming decade economic research into petroleum economics must make a special effort to combine economic with geological considerations. This means that the importance of the reserve/production ratio must be recognised. In this short appendix I would like to present a new fundamental result relating to the reserve-production ratio. Important work on this topic is now being done by Leo Drollas of British Petroleum and Pierre DesPrairies and his associates at the French Institute of Petroleum. There is also a recognition of this issue in research being carried out by the OPEC secretariat.

In what follows R is reserves at the beginning of a period, Q is the production of oil in the given period, and D the increase in reserves during a given period due to e.g. exploration. Thus we can write for reserves at time T:

$$R(T) = R(O) - \int_0^T Q(O)e^{nt}dt + \int_0^T D(t)dt$$

Note here that $Q(O)e^{nt} = Q(t)$, where n is the constant rate of growth of production. I also assume a constant rate of growth of reserves g, and thus we have:

$$g = \frac{D(t)}{R(t)}$$

By using this expression, and by dividing the first equation by Q(T) we get:

$$\theta(T) = \frac{R(O)}{Q(T)} - \frac{1}{Q(T)} \int_0^T Q(t)dt + \frac{1}{Q(T)} \int_0^T gR(t)dt$$

Now, noting that $(Q)T = Q(0)e^{nT}$, and thus we get via differentiation $Q'(T) = nQ(T)$, we can differentiate the expression for $\theta(T)$ with respect to T to get:

$$\frac{d\theta}{dT} = -\frac{nR(O)}{Q(T)} + \frac{n}{Q(T)} \int_0^T Q(t)dt - \frac{1}{Q(T)} Q(T)$$
$$- \frac{n}{Q(T)} \int_0^T gR(t)dt + \frac{1}{Q(T)} gR(T)$$

This can be simplified right away to:

$$\frac{d\theta}{dT} = (g - n)\theta(T) - 1$$

This expression is revealing. Assume for example that $n = 0$ and $g = $ five per cent, where growth is measured in terms of the reserve base at the beginning of the year. Intuitively, we might jump to the conclusion that the R/Q ratio is growing, but this need not be so. In fact, what the above expression says is that for the R/Q ratio to be growing, we must have an R/Q ratio greater than 20 at time T. Just how can this be so?

The answer is that in this exercise the growth of reserves is measured from the amount existing at the beginning of each year. For example, if reserves at the beginning of the year are 150, and the annual reserve growth is five per cent, then reserves *ceteris paribus* increase by 7.5. But reserves at the end of the period are not

150 + 7.5, but 150 + 7.5 minus production for that year. Thus, as indicated in the above algebra, for total reserves to expand over time, reserve additions must be large relative to consumption. If there is some doubt in the reader's mind about this result, the best way to examine this issue is to construct a simple numerical example.

Bibliography

Adelman M.A., *The supply and price of natural gas*, Basil Blackwell, Oxford (1962).
Alchian A., Costs and Outputs, in *The allocation of resources*, edited by M. Abramovitz, Stanford University Press, Palo Alto (1959).
Banks F.E., *The political economy of oil*, D.C. Heath & Co, Lexington & Toronto (1980).
────── *Soviet natural gas and the Western European energy shortage: the solution is the problem*, paper presented at a conference of the IAEE, Cambridge University (1982).
────── European reliance on Soviet gas exports: a comment, *The Energy Journal* (July 1983).
────── The political economy of European natural gas, *Chemical Economy and Engineering Review* (May 1984).
────── *The political economy of coal*, D.C. Heath & Co, Lexington & Toronto (1985).
────── Natural gas in Australia, *Energy Policy* (February 1986).
Bonfiglioli G., Economics of gas imports to Europe from OAPEC, *Oil and Gas Journal* (4 August 1980).
Boucher J. and Smeers Y., Gas trade in the European Community during the seventies, *Energy Economics* (February 1985).
────── Measuring the consequence of the take-or-play clause in natural-gas consuming countries, in *Macroeconomic prospects for a small oil-exporting country*, edited by O. Bjerkholt and E. Offerdal, Martinus Nijhoff Publishers, Dordrecht (1985).
Bourcier P., Julius D., Moulin P. and Palmer K., The economics of natural gas development, *Energy* (April 1985).
Broadman H.G. and Montgomery W., *Natural gas markets after deregulation*, Resources for the Future, Washington DC (1983).
Campbell R.W., *Trends in Soviet oil and gas industry*, Johns Hopkins Press, Washington DC (1976).
Chenery H.B., Engineering production functions, *Quarterly Journal of Economics* (November 1949).
Choe B.J., *The outlook for thermal coal*, World Bank staff commodity working papers, No. 12 (1985).
Czerniejewicz W., Natural gas in the Federal Republic of Germany: facts and prospects, in *Energy and Economy: Global Interdependencies* (vol. 2), edited by Mark Baier, paper presented at the Seventh International Conference of the IAEE, Bonn (June 1985).
Dahl C. and Boyd R., The effects of the Soviet gas pipeline on Western European energy markets: econometric estimation, in *Energy and Economy: Global Interdependencies* (vol. 2), edited by Mark Baier, paper presented at the Seventh International Conference of the IAEE, Bonn (June 1985).
Dam K., The pricing of North Sea gas in Great Britain, *The Journal of Law and Economics* (April 1970).

Davis J.D., *Blue gold: the political economy of natural gas*, George Allen and Unwin, London (1984).
Davis T.A., *The outlook for natural gas in Western Europe to 2000*, stencil (1984).
Drayton G., *The market for LPG in the 1980s*, Economist Intelligence Unit, special report no. 961 (1981).
Erdman P., *Paul Erdman's money guide*, Secker and Warburg, London (1984).
Estrada J., *The importance of the natural gas industry in the Soviet Union*, working paper, Bedriftsøkonomisk Institutt, Oslo (1984).
Fesharaki F. and Hoffman S., *OPEC natural gas: reserves, utilization and trade*, stencil, East-West Center (1983).
Frazer F., *Gas prospects in Western Europe*, Financial Times business information, London (1982).
Fugelberg O., *Demand for natural gas in Western Europe: the case of France*, working paper, Bedriftsøkonomisk Institutt, Oslo (1984).
Gray S., New Zealand switches to natural gas, *The Financial Times* (April 11 1984).
Gregory R.G., Some implications of the growth of the mineral sector, *Australian Journal of Agricultural Economics* (August 1976).
Hawdon D., *The economics of the natural gas market and its competitiveness*, stencil, University of Surrey (1985).
Hodges R.E., P.C. program selects gas-line sizes, *Oil and Gas Journal* (April 1985).
Hough G. Vernon, LNG market: sales increase as price weakens, *The Petroleum Economist* (December 1984).
―――― Natural gas, *Quarterly Energy Review* (October 1985).
Hubbard G. and Weiner R., *Regulation and bilateral monopoly: long term contracting in natural gas*, discussion paper series, John F. Kennedy School of Government, Harvard University (1984).
Karplus R.S., *Competitiveness of Norwegian and Soviet gas supplies*, stencil (1985).
Khan A.R., US natural gas at the crossroads, *OPEC Quarterly Review* (spring 1984).
Lemon J.R., Competition among Canadian and United States producers for the US natural gas market, in *Energy and Economy: Global Interdependencies* (Vol. 2), edited by Mark Baier, paper presented at the Seventh International Conference of the IAEE, Bonn (June 1985).
Lesourd J.B., Percebois, J. and Ruiz J.M., *'Equilibre et deséquilibre sur le marche international du gaz naturel*, stencil, University of Grenoble (1986).
Lorentzon L. and Roland K., *Norway's export of natural gas to the European gas market: policy issues and model tools*, stencil, Oslo (1985).
―――― Roland K. and Asheim A., *Cost structure and profitability of a north sea oil and gas field*, stencil, Oslo (1986).
MacAvoy P.W., *Price formation in natural gas fields*, Yale University Press, New Haven (1962).
Manne A.S., Roland K. and Stephan G., The Western European market for natural gas: security of supplies, *Energy Policy* (February 1986).
Mathiesen L., Roland K. and Thonstad K., *The European natural gas*

market: degrees of market power on the selling side, stencil, Oslo (1986).

Mindermann F., Natural gas in West European energy supplies, *The Journal of Energy and Development*, no.1 (1984).

Molle W. and van der Vlies J., Long term changes in the energy markets of the European Community: patterns and causes, in *Energy and Economy: Global Interdependencies* (vol. 2), edited by Mark Baier, paper presented at the Seventh International Conference of the IAEE, Bonn (June 1985).

Nemetz P.N. and Vertinsky I.B., Japan and the international market for LNG, *Columbia Journal of World Business* (spring 1984).

Newman P. and Brotherton P., North West shelf development, in *Energy Policy in Western Australia*, edited by Frank Harman and Peter Newman, Murdoch University Press, Murdoch, Western Australia (1984).

Noreng O., *Energy policy and prospects in Norway*, working paper, Bedriftsøkonomisk Institutt, Oslo (1986).

Paulette C.H., A new approach to use of the revised panhandle formulas, *The Pipeline Engineer* (March 1968).

Peebles M.W., *Evolution of the natural gas industry*, MacMillan, London (1980).

────── *Prospects for natural gas through to the year 2000*, paper presented at the Sixteenth World Gas Conference, Munich (24–7 June 1985).

Percebois J., Gas market prospects and relationship with oil prices, *Energy Policy* (August 1986).

Roland K., *Natural gas supply and demand in Western Europe, 1990 and 2000*, International Energy Program Report, Stanford University (February 1984).

────── Hvordan selge Norsk gass etter at Sleipneravtalen strandet? *Sosialøkonomen* (no. 8 1985).

Russell J., *Geopolitics of natural gas*, Ballinger, Cambridge, Massachusetts (1983).

Schramm G., The changing world of natural gas utilization, *Natural Resource Journal* (April 1984).

Smith V., *Investment and production*, Harvard University Press, Cambridge, Massachusetts (1961).

Steenblik R., *The price of coal*, Town and Country Planning Association, London (September 1983).

────── *International coal supply and the bargaining power of the developing countries*, stencil (1985).

Stern J.P., *Soviet Natural Gas in the World Economy*, Association of American Geographers, Washington DC (1979).

────── *Soviet natural gas development to 1990*, D.C. Heath & Co, Lexington and Toronto (1981).

────── *Natural gas trade in Europe — the policies of exporting and importing countries*, Heinemann Educational Books, London (1984).

────── *Natural gas trade in North America and Asia*, Gower, Aldershot (1985).

────── After Sleipner: a policy for UK gas supplies, *Energy Policy* (February 1986).

Tussing A.R. and Barlow C., *The natural gas industry*, Harper and Rowe, London (1984).
Uhler R., The supply of natural gas reserves in Alberta, in *Progress in Natural Resource Economics*, edited by A. Scott, Clarendon Press, Oxford (1985).
Watkins G.C. and Waverman L., *Canadian natural gas export pricing behavior*, stencil, University of Calgary (1985).
Waverman L., *Natural gas and national policy*, Toronto University Press, Toronto (1973).
Weale G. and Pariente-David S., European natural gas markets: prospects and risks, in *Energy and Economy: Global Interdependencies* (vol. 2) edited by Mark Baier, paper presented at the Seventh International Conference of the IAEE, Bonn, June (1985).

Index

Abu Dhabi 126
Adelman, M. 26
Airlines 176
Alchian, A. 153, 169
Algeria 13, 34, 47, 109, 119, 129
American Gas Association 64
Ammonia 103
Associated gas 3, 4, 42
Australia 22, 26, 30, 36, 106, 159

Belgium 121
Berzelius, Jöns Jacob 5
Border Gas 73
Bulk fuels 62, 65

Canada 4, 30, 38, 45, 71, 159
Canadian Energy Board 72, 108
Chenery, H. 153
Chernobyl 94
China 109
Choe, B.J. 29
City Gate Contracts 68
Coal 22, 64, 89
Colliers 27
Compressed natural gas 38, 102
Compressor equipment 42, 78
Conversion factors 6
Czerniejewicz, W. 139

D'Avignon plan 79
Davis, J. 130, 136
Debt 179
Denmark 123, 141
Depletion (of gas fields) 153
DesPrairies, P. 191
Drilling costs 59, 60
Drollas, L. 191
Dual (fuel use) capability 63, 138

Egypt 101
El Paso Gas 129

Finland 123
France 121, 137
Fugelberg, O. 153
Futures contracts 182
Futures exchanges
  delivery 185
  margin 185
Futures markets 181
  hedge 182

Game theory 87
Garnaut, T. 113
Gas turbines 103
Gasol (bottled gas) 4
Gaz de France 129
Gorbachev, M. 93
Gregory, R. 114
Groningen 2

Hawdon, D. 153, 163
Helmont, Jan van 5
Hodges, R.E. 165
Hofman, Wilhelm von 5
Hubbard, G. 69

Indexing (formula) 82, 124, 130
International Association of
  Energy Economists (IAEE) 1
International Energy Agency
  (IEA) 109, 120
Iran 9, 14
Italy 122

Japan 11, 72, 79, 96, 105, 176
Joint ventures 57

Khan, A.R. 153, 156
Kuwait 144

Labour markets 177
Liberman reforms 77
Liquefied natural gas (LNG) 44, 47, 106, 110, 126, 135

Liquefied petroleum gas (LPG) 4, 102

MacAvoy, P. 157
Macroeconomic markets, equilibrium in 175
Malaysia 114
Manne, A. 149
Marginal cost 169
Metallurgical coal 23
Methane 4
Methanol 102
Mexico 71
Minimum billing 131
Ministry of International Trade and Industry (MITI) 30
Mitterand, F. 81, 138

National Energy Board (of Canada) 117
National Gas Act (UK) 135
Natural gas
  liquids 4, 96
  offshore 6
  reserves 9
  storage 112, 132
  unconventional 4
Natural Gas Policy Act 68, 74
Netback 51
Netherlands 33, 34, 122, 142
New York Merchantile Exchange (NYMEX) 183
New Zealand 115
Non-associated gas 3
Norway 34, 56, 88, 144, 159

OECD 76, 96
Oil 16
  price 179
Open interest 184
Organisation of Petroleum Exporting Countries (OPEC) 16, 35, 36, 188

Pariente-David, S. 124
Paulette, C.H. 163
PEMEX (The Mexican National Oil Company) 71
Percebois, J. 13

Petrochemicals 20, 176
Pipeline companies 61, 67, 70, 71
Pipelines 42, 58, 78, 89, 119, 154
  average costs 169
Primary energy 2

Qatar 126
Qatar Liquefied Natural Gas Co. (QALIGAS) 99

Reagan, R. 80, 89, 147
Regulation and deregulation 66
Reserve-Production ratio 19, 39, 60, 61, 190
Resources rent tax 113
Roland, K. 149, 153

Saudi Arabia 91, 96, 144, 188
Saudi Arabian Basic Industries Corporation (SABIC) 98
Scanlan, T. 18
Secondary energy 2
Shipping (LNG) 118
Short run oil pricing 190
Sleipner gas field 136
Smith, V. 153
Soviet Union 9, 33, 44, 76, 106, 119
Spain 123
Spallanzani, L. 5
Spot markets 64
Stanford Institute for Mathematical Studies in the Social Sciences 150
Statoil (Den Norske Stats Oljeselskap) 140
Steenblik, R. 31
Stephan, G. 149
Stern, J.F. 76, 149
Stockmarkets (or sharemarkets) 189
Sweden 79, 95, 104, 123, 159

Take-or-pay clause 131
Tankers 118
Technology, change and investment 178

INDEX

Third World 175
Town gas 5
TransMed pipeline 97, 141
Troll gas field 126, 140

Unemployment 76
United Arab Emirates 96, 144, 188
United Kingdom 122, 134
United States 14, 37, 58, 80, 155, 176
United States army 90
United States Geological Survey 17

United States Natural Gas Clearinghouse 65
Units 6

Venezuela 4

Watkins, G.C. 118
Waverman, L. 118
Weale, G. 124
Weiner, R. 69
West Germany 121
World Bank 100
World Coal Report (Wocol) 23